Computer-Aided Materials Selection During Structural Design

Committee on
Application of Expert Systems to
Materials Selection During Structural Design

National Materials Advisory Board
Commission on Engineering and
Technical Systems
National Research Council

NMAB-467

National Academy Press
Washington, D.C. 1995

NOTICE: The project that is the subject of this report was approved by the Governing Board of the National Research Council, whose members are drawn from the councils of the National Academy of Sciences, the National Academy of Engineering, and the Institute of Medicine. The members of the committee responsible for the report were chosen for their special competencies and with regard for appropriate balance.

This report has been reviewed by a group other than the authors according to procedures approved by a Report Review Committee consisting of members of the National Academy of Sciences, the National Academy of Engineering, and the Institute of Medicine.

This study by the National Materials Advisory Board was conducted under ARPA Order No. 8475 issued by DARPA/CMO under Contract No. MDA 972-92-C-0028 with the U.S. Department of Defense and the National Aeronautics and Space Administration.

The views and conclusions contained in this document are those of the authors and should not be interpreted as representing the official policies, either expressed or implied, of the Defense Advanced Research Projects Agency or the U.S. Government.

Library of Congress Catalog Card Number 94-69233
International Standard Book Number 0-309-05193-2

Available in limited supply from:
National Materials Advisory Board
2101 Constitution Avenue, NW
Washington, D.C. 20418
202-334-3505 nmab@nas.edu

Additional copies are available for sale from:
National Academy Press
2101 Constitution Avenue, NW
Box 285
Washington, D.C. 20055
800-624-6242
202-334-3313 (in the Washington Metropolitan Area)

Copyright 1995 by the National Academy of Sciences. All rights reserved.

Printed in the United States of America.

Abstract

The selection of the proper materials for a structural component is a critical engineering activity. It is governed by many, often conflicting factors that typically include service requirements, design life, materials availability, database accessibility, manufacturing constraints, repair and replacement strategies, client preferences, and cost. The incorporation of computer-aided materials selection systems into computer-aided design and computer-aided manufacturing operations could assist designers by suggesting potential manufacturing processes for particular products to facilitate concurrent engineering, recommending various materials for a specific part based on a given set of characteristics, or proposing possible modifications of a design if suitable materials for a particular part do not exist. This report reviews the structural design process, determines the elements and capabilities required for a computer-aided materials selection system to assist design engineers, and recommends the research and development areas of materials database, knowledge base, andmodeling required to develop a computer-aided materials selection system.

THE NATIONAL ACADEMIES

National Academy of Sciences
National Academy of Engineering
Institute of Medicine
National Research Council

The **National Academy of Sciences** is a private, nonprofit, self-perpetuating society of distinguished scholars engaged in scientific and engineering research, dedicated to the furtherance of science and technology and to their use for the general welfare. Upon the authority of the charter granted to it by the Congress in 1863, the Academy has a mandate that requires it to advise the federal government on scientific and technical matters. Dr. Bruce Alberts is president of the National Academy of Sciences.

The **National Academy of Engineering** was established in 1964, under the charter of the National Academy of Sciences, as a parallel organization of outstanding engineers. It is autonomous in its administration and in the selection of its members, sharing with the National Academy of Sciences the responsibility for advising the federal government. The National Academy of Engineering also sponsors engineering programs aimed at meeting national needs, encourages education and research, and recognizes the superior achievements of engineers. Dr. Robert M. White is president of the National Academy of Engineering.

The **Institute of Medicine** was established in 1970 by the National Academy of Sciences to secure the services of eminent members of appropriate professions in the examination of policy matters pertaining to the health of the public. The Institute acts under the responsibility given to the National Academy of Sciences by its congressional charter to be an advisor to the federal government and, upon its own initiative, to identify issues of medical care, research, and education. Dr. Kenneth I. Shine is president of the Institute of Medicine.

The **National Research Council** was organized by the National Academy of Sciences in 1916 to associate the broad community of science and technology with the Academy's purposes of furthering knowledge and advising the federal government. Functioning in accordance with general policies determined by the Academy, the Council has become the principal operating agency of both the National Academy of Sciences and the National Academy of Engineering in providing services to the government, the public, and the scientific and engineering communities. The Council is administered jointly by both Academies and the Institute of Medicine. Dr. Bruce Alberts and Dr. Robert M. White are chairman and vice chairman, respectively, of the National Research Council.

www.national-academies.org

Committee on Application of Expert Systems to Materials Selection During Structural Design

FRANK W. CROSSMAN *Chair*, Director, Material Sciences, Lockheed Palo Alto Research Laboratory, Palo Alto, California
JAN D. ACHENBACH, Director, Center for Quality Engineering and Failure Prevention, Northwestern University, Evanston, Illinois
HAROLD L. GEGEL, Director, Processing Science Division, Universal Energy Systems, Dayton, Ohio
RICHARD N. HADCOCK, Vice President, RNH Associates, Inc., Huntington, New York
THOMAS S. KACZMAREK, North American Operation Manufacturing Center, General Motors Corporation, Warren, Michigan
J. GILBERT KAUFMAN, Vice President, Technology, The Aluminum Association, Washington, D.C.
MICHAEL ORTIZ, Engineering Department, Brown University, Providence, Rhode Island
FRIEDRICH B. PRINZ, Rodney H. Adams Professor of Engineering, Departments of Mechanical Engineering and Materials Science, Stanford University, Stanford, California
JAN SCHREURS, Westinghouse Science and Technology Center, Westinghouse Electric Company, Pittsburgh, Pennsylvania
VOLKER WEISS, Professor of Engineering and Physics, Chairman, Department of Mechanical Aerospace and Manufacturing Engineering, Syracuse University, Syracuse, New York

LIAISON REPRESENTATIVES

RALPH P. I. ADLER, Chief, Metals Research Branch, Army Research Laboratory—Materials Directorate, Watertown, Massachusetts
WILLIAM BARKER, Defense Sciences Office, Defense Advanced Research Projects Agency, Arlington, Virginia
ANDREW CROWSON, Director, Metallurgy and Materials Science Division, Research Triangle Park, North Carolina
WALTER M. GRIFFITH, Deputy Director, Metals and Ceramics Division, Materials Directorate, Wright Patterson Air Force Base, Ohio
CRAIG MADDEN, Research Engineering Group, David Taylor Research Center, Bethesda, Maryland
RONALD G. MUNRO, Physicist, Ceramics Division, National Institute of Standards and Technology, Gaithersburg, Maryland

NMAB STAFF

ROBERT M. EHRENREICH, Senior Staff Officer
PAT WILLIAMS, Senior Secretary

Acknowledgments

The committee would like to express its appreciation to the following individuals for their presentations to the committee: J. Hendrix of Hercules Incorporated, O. Richmond of ALCOA, and D. Marinaro of PDA Engineering. The committee would also like to thank Larry Ilcewicz for hosting a site visit to the Boeing facility in Seattle, Washington, and the following individuals for their presentations: H. Shomber, A. Falco, T. Lackey, T. Richardson, B. Das, B. Backman, A. Miller, J. Boose, P. Rimbos, and G. Swanson. The committee acknowledges with thanks the contributions of Robert M. Ehrenreich, Senior Staff Officer, and Pat Williams, Senior Secretary, to the project.

Preface

The Department of Defense and the National Aeronautics and Space Administration requested that the National Materials Advisory Board convene a committee to study the application of expert systems to materials selection during structural design. The objectives of the study were to determine (1) the components needed for an effective computer-assisted, concurrent engineering design system, (2) the barriers preventing the development of such a system, and (3) the research and development required to construct such a system.

The committee met six times between June 1991 and September 1992. The first meeting focused on developing a perspective on the study scope and an approach for assessing the primary, underlying technologies pertinent to the study *via case studies and an industry site visit.*[1] The second meeting provided an opportunity for each committee and liaison member to describe his or her research and technical experience pertinent to the committee charter. Committee members also presented case studies of computer-aided materials selection systems with which they had experience. This helped determine the state of the art of such systems and provide examples of design decisions involving geometric relations; design rules associated with performance, processing, manufacturing, and supportability; and advanced computer concepts and technologies that aid the design optimization process. The committee then held three study sessions focused on product design, materials supply and development, and state-of-the-art computer-aided systems technology for materials selection. The first study session consisted of a site visit to the Boeing Commercial Airplane Company in Seattle, Washington, to learn about materials selection within the airplane design process. The second session consisted of presentations by representatives of ALCOA and Hercules on materials modeling, certification, and the supplier-designer interaction. The third session focused on computer demonstrations of some state-of-the-art systems that aid the materials selection process, to determine current capabilities and identify barriers to the development of an optimal system.

This report is divided into five chapters. Chapter 1 defines the study scope and committee charge. Chapter 2 places materials selection within the context of the design process, using the Boeing Commercial Airplane Group as a case study. Chapter 3 presents the committee's vision of a full-function, computer-aided materials selection system. Chapter 4 reviews the underlying information technologies pertinent to the materials selection process, determined by the examination of the case studies listed in Appendix B. Chapter 5 discusses the issues preventing the development of computer-aided materials selection systems and outlines the recommendations for the research and development required to attain the envisioned system. The appendices include (1) a glossary of acronyms, (2)

---

[1] The number of computer-aided systems on the market is rapidly growing, with new products being introduced daily (see Schorr and Rappaport, 1989; Rappaport and Smith, 1991; Smith and Scott, 1991). Since any compilation of systems would be rapidly out-of-date, the committee determined the current capabilities of computer-aided systems by examining the 30 case studies listed in Appendix B.

a complete list of the case studies reviewed by the committee, (3) a review of some of the knowledge-representation tools and technologies discussed in the report, and (4) two examples of the case studies reviewed by the committee that typify the materials selection and database systems currently available.

Comments or suggestions that readers of this report wish to make can be sent via Internet electronic mail to nmab@nas.edu or by FAX to the National Materials Advisory Board at 202/334-3718.

Frank W. Crossman, *Chair*

Contents

	EXECUTIVE SUMMARY	1
	Vision of a Computer-Aided Materials Selection System	1
	Strategies for Overcoming Barriers	2
	General Conclusions and Recommendations	4
1	INTRODUCTION	7
	Benefits	7
	Definitions	8
	Study Objectives and Scope	8
2	MATERIALS SELECTION IN STRUCTURAL DESIGN	11
	Concurrent Engineering and Design Organization	11
	Materials Selection During Computer-Aided Design	13
	Summary of Materials Information Requirements in Design	15
3	ENHANCING THE MATERIALS SELECTION PROCESS IN DESIGN: A VISION	19
	Integrated Engineering Support in Integrated Enterprises	19
	Supporting Strategic Material Decisions	20
	Supporting Routine Material Decisions	21
	Supporting Innovative Materials Selection in Design	21
	Summary	22
4	INFORMATION TECHNOLOGIES PERTINENT TO THE MATERIALS SELECTION PROCESS	25
	Databases and Knowledge Bases	25
	Modeling and Analysis Systems	29
5	CONCLUSIONS AND RECOMMENDATIONS	37
	Strategies for Overcoming Barriers	37
	General Conclusions and Recommendations	39
	References	41
Appendix A:	Glossary of Acronyms	45
Appendix B:	Case Studies Reviewed by the Committee	47

Appendix C:	Review of Selected Knowledge-Representation Techniques and Tools	49
Appendix D:	Knowledge-Based Integrated Design System	57
Appendix E:	An Intelligent Knowledge System for Selection of Materials for Critical Aerospace Applications	63
Appendix F:	Biographical Sketches of Committee Members	69

List of Figures

2-1	The sequential engineering approach to structural design	11
2-2	A typical comparison of sequential and concurrent engineering	12
2-3	An example of a typical structures DBT hierarchy: Boeing 777 horizontal stabilizer DBTs	12
2-4	The expertise in a structural design concurrent engineering team	13
2-5	The interactions of a typical DBT	13
2-6	A model of the wing of the Grumman X-29 and associated finite element analysis	14
2-7	Structures and materials design interactions	15
3-1	The conceptual architecture of a Computer-Aided Materials Selection System	24
4-1	Automobile side marker	30
4-2	Concurrent engineering environment including inspectability	33
4-3	POD curves for two scanning plans	35
C-1	The differences between wireframe, surface model, and solid model representational domains	52
C-2	An example of the LOOS system to define the structure or "topology" of a layout	55
D-1	Information flow in IPD	58
D-2	Methods developers' frame of reference	59
D-3	Control flow between roles in the blade design assistant	60
D-4	Blade design assistant	61
E-1	Model of an intelligent knowledge system applicable to the material selection problem	64
E-2	Specific system architecture for the prototype IKSMAT	64

List of Tables

2-1	Examples of Materials Information Required During Product Design	16
2-2	Typical Product Design Requirements for Aircraft Structure Development	16
2-3	Summary of Designer "Wants"	17
3-1	Summary of the Materials-Specific Information Technologies and Some of the Primary Computer Technologies Required for a Computer-Aided Materials Selection System	23
4-1	Representative Applications Based on Knowledge of Materials	27
4-2	Steps in the Development of a Process Model	31

Computer-Aided Materials Selection During Structural Design

Executive Summary

Selecting the proper materials for a structural component is critical to engineering design. Materials selection is governed by many factors, some of which are in opposition. The principal selection factors include the service requirements and design life of the product; the availability of candidate materials and the appropriate data on application-specific properties for them; the company's *make* or *buy* decision for the system components; the customer preferences; and, most importantly, the total life-cycle cost.

The use of computer-aided systems could reduce cost and design rework and requalification by providing engineering design teams with the most current materials-property data, knowledge of factors such as materials options and life-cycle costs, and available materials for a design based on experience derived from previous product developments. A Computer-Aided Materials Selection System (CAMSS) with learning capabilities would also ensure the proper archiving of materials selection decisions for future reference and accelerate the application of new materials and processing technologies by providing designers with an expanded range of possible materials and manufacturing methods for a given set of product characteristics and cost-performance criteria.

This study concentrated on the materials-specific knowledge elements of a computer-aided system. The Committee on the Application of Expert Systems to Materials Selection during Structural Design determined that the development of generic computer-aided systems is already receiving a great deal of attention within the computer science and engineering community. The basic information requirements for a computer-aided system for materials selection are receiving little attention within the materials community, however. Thus, the committee assessed that this study would have the greatest impact if it (1) detailed the capabilities required for computer-aided systems to be of value to the materials selection process during concurrent engineering; (2) identified the issues inhibiting the development of such a system; and (3) recommended materials-specific applications and developments in database, knowledge base, and materials modeling that would aid the production of a knowledge element appropriate for computer-aided systems for materials selection during design.

VISION OF A COMPUTER-AIDED MATERIALS SELECTION SYSTEM

The committee developed a conceptual architecture for a CAMSS that depicts the supporting materials-specific information technologies required. The objective of a CAMSS should be to provide design options for consideration by the design engineering team. Design and materials advisor tools should be available throughout the concurrent engineering process. Significant material properties as well as emerging considerations, such as processing and product recycling costs, will be increasingly supported by the information infrastructure. Materials knowledge should be made accessible to the engineer as reference data through design advisors that interact with product and process models that analyze, critique, improve, or optimize the design.

Major tools in the integrated environment will provide the following materials-specific capabilities: managing electronic repositories of data and documents, searching past development histories to find similar or analogous products, managing requirements, analyzing performance characteristics, modeling manufacturing and maintenance characteristics, estimating costs, suggesting improvements to the proposed product or process description, and storing the rationale for materials selection decisions for future reference. The alternative selected during concept evaluation would then be available for further refinement by the designer in a coarse-to-fine development process. To accomplish this, the CAMSS should make use of available computer-aided systems technologies. Computer-aided systems consisting of both heuristic and quantifiable design rules can be developed for subsets of the design knowledge base.

Computing technology no longer presents a barrier to the development of a CAMSS. The wide range of both hardware and software capabilities is rapidly reducing the cost of representing and implementing computer-aided system logic and process simulation within affordable limits. Advances in reduced instruction set computer (RISC) chip technology allow inexpensive workstations to perform both design layout and an embedded structural analysis or materials processing simulation. Visualization techniques coupled with simulation of system behavior at many levels can be a powerful means of conveying information to design team members. The continued evolution of cost-effective, high-performance computing in conjunction with a national information superhighway infrastructure will further assist the nation's manufacturing sector in becoming more competitive in the international market-place.

STRATEGIES FOR OVERCOMING BARRIERS

The committee identified two main areas that are currently preventing the development of a CAMSS: (1) database and knowledge base design, implementation, instantiation, and management and (2) structural design modeling technologies.

Database and Knowledge-Base Barriers

The design, implementation, instantiation, and maintenance of materials properties databases and knowledge bases are integral to the development of an effective CAMSS. For example, a design engineer cannot use a system if the underlying databases contain obsolete, extraneous, unverified, or incomplete information. The committee has found that the database and knowledge base area is currently inhibited by five barriers.

1. *Standardization of databases and knowledge bases*—Constructing databases and knowledge bases that contain the relevant information required for the design process and developing systems that locate and present this data are two difficult problems because of the amount of extraneous information available and the lack of standards in the content of databases and knowledge bases. *To overcome these barriers, the committee recommends that (1) standards and guidelines be developed for electronic data quality, capture, storage, analysis, and exchange (following the Computer-Aided Acquisition and Logistics Support and the Standard for the Exchange of Product approaches) and knowledge base content and construction; (2) CAMSS be designed to accept a variety of database taxonomies through the use of active, "intelligent" data dictionaries that aid the identification and conversion of the contents of different databases for use in the system; (3) links between materials databases and knowledge bases be improved and computer networks for materials-specific information communication be created (e.g., an electronic* Journal of Materials Selection in Structural Design, *a national materials bulletin board on Internet, or a linked network of worldwide materials data systems); and (4) electronic technical assistance be provided to design teams in electronic formats.*

2. *Status of knowledge capture*—Methods for knowledge capture are required to enhance the lessons-learned segment of CAMSS. These include establishing knowledge-representation taxonomies, technical context standards, and techniques to update and access this information rapidly. *To overcome this barrier, the committee recommends that (1) materials and computer scientists collaborate in the development of suitable knowledge-capture systems for use in CAMSS; (2) industry design teams be encouraged to establish electronic technical databases by electronic capture of all design discussions, decisions, and lessons learned in free text, spread-sheet, computer-aided design (CAD) standards, and other multimedia formats; and (3) industry design teams be encouraged to assign specific functions within the team to specialize, categorize, index, and filter the accumulated design knowledge base and locate and access other design knowledge bases.*

3. *Diffuse responsibility for generating databases*—The issue of how to coordinate materials developers, component users, and materials societies to generate and integrate materials property databases requires resolution. Materials suppliers predominantly leave materials qualification programs to the user because of concerns that they will be held liable for system malfunctions caused by failures and that users will only employ

materials that they themselves have qualified. Materials societies generally do not have the resources necessary for large projects. Component manufacturers typically only qualify materials for a given application and treat the data as proprietary. *To overcome this barrier, the committee recommends that (1) national team efforts of users, suppliers, materials societies, and standards organizations develop integrated material qualification programs that relate to design requirements and eventual use and (2) the resultant appropriate, independently verified data be made available in a national information infrastructure to provide a realistic, initial appraisal of the advantages of a material.*

4. *Disclosure of materials data*—In general, companies protect as proprietary all databases and knowledge bases that contain materials properties and production-related data, such as (1) state-of-the-art information, projections, or forecasts; (2) manufacturing labor standards, rates, and price data; and (3) weight, performance, and cost tradeoff data and criteria. *To overcome this barrier, the committee recommends that CAMSS be designed to assure that proprietary portions of databases and knowledge bases are fully protected.*

5. *Investment to maintain databases*—It is important that the information within a database be constantly monitored, verified, and updated to ensure that the best possible information is available. *To overcome this barrier, organizations must (1) assign the responsibility for maintenance of databases to a centralized function, such as a data administrator with domain experts identified to act as curators of the knowledge base, and (2) provide long-term support for database maintenance once the program is established.*

Structural Design Modeling Technology Barriers

Modeling in structural design will be an important component of any CAMSS both to provide valid details on which to base tradeoff decisions and to reduce reliance only on materials-properties databases. Modeling techniques are required for geometric reasoning, material responses on multiple scale levels, materials processing, manufacturing processing performance, product performance, and life-cycle issues such as inspectability. Modeling techniques will also be required that simulate new materials by successive extrapolation from the properties of existing materials or by calculation from first principles. The committee identified two barriers to the development of modeling.

1. *Optimization modeling*—As opposed to simply showing tradeoffs between design parameters input by users, modeling techniques will be required that can suggest modifications to optimize designs and manufacturing processes. Process optimization is an important ingredient of integrated product-process design and will be used more and more in the future as the industry fully adopts concurrent engineering to reduce manufacturing costs and converge on manufacturing solutions in a shorter time. To be useful, modeling must also be done rapidly and accurately, using normal design parameters and information from multiple knowledge bases. If modeling techniques are too slow, untrustworthy, or unable to access the proper information, they will languish. *To overcome these barriers, the committee recommends that (1) materials scientists and computer engineers from industry and university collaborate to develop advanced modeling techniques to reduce reliance on straight materials data, introduce expert knowledge, provide a credible basis for tradeoff decisions, and increase trust in CAMSS; and (2) materials scientists participate in basic and applied research that establishes links between materials models at several scales (e.g., atomic, molecular-crystal, cluster-grain size, polycrystal-aggregate, substructure, structure, and system).*

2. *Cultural and educational barriers to implementing modeling and analysis technology*—The design process is traditionally a heuristic trial-and-error approach. Increased reliance on modeling techniques requires establishing confidence that the improved design solutions can be developed in a shorter time period. Current engineering programs do not stress the importance of training in either materials synthesis and processing or computer science. For modeling and

analysis systems in a CAMSS to be useful and effective, future engineers must receive training in computer systems, modeling and analysis systems theory, and their application to the design process. *To overcome the cultural and educational barriers, the committee recommends that institutions of higher learning develop interdisciplinary programs led jointly by experts in materials science and engineering, design, and computer science that (1) expose student teams to basic approaches to computer-assisted concurrent engineering design systems in order to produce knowledgeable workers with a broad understanding of the science of processing, (2) train journeymen or master technologists to use this new technology to push acceptance of process modeling techniques to the shop floor, and (3) encourage younger faculty members to collaborate with colleagues in other departments (e.g., materials science, the traditional engineering fields, and computer science) on interdisciplinary design projects and computer-assisted concurrent engineering design systems.*

GENERAL CONCLUSIONS AND RECOMMENDATIONS

The areas inhibiting the development and implementation of CAMSS discussed above can only be overcome by a multipronged initiative with full participation and support by the integrated product development teams (IPDTs) and materials and computer scientists and engineers in the government research and development (R&D) agencies, universities, and industrial organizations. The implementation of this vision will require (1) the development of significant demonstrations of CAMSS and disseminating the results; (2) the continued expansion of electronic storage of materials information; (3) the rapid adoption and application of developing methods of computer science and technology to enhance the representation of materials design knowledge; (4) the continued development of multilevel (atomistic to macroscopic) materials processing and constitutive behavior models that reliably predict performance and manufacturability at the scale of application; and (5) the implementation of methods to address inspectability, reliability, and maintainability.

Adherence to uniform computing and materials description standards in such programs is essential to the networked linking of individual tools into much larger design knowledge and support systems in the future. The committee recommends a higher level of communication among materials-specific information systems researchers and developers through a more formal electronic interchange of research information, network-linked use of computer-aided system tools, and access to electronic materials knowledge bases.

Recommendations specific to developers and users of CAMSS are

- *Government policy makers* should promote (1) the development of pre-competitive R&D programs that encourage industry, university, and government laboratories to leverage expertise and knowledge to reduce the time to develop, standardize, and implement product design support systems and materials-specific information technologies and (2) the use of the information superhighway as a means for expediting the sharing of technical information and memory among federal agencies, industries, and materials societies.
- *Government R&D organizations* (Department of Defense, Advanced Research Projects Agency, National Aeronautics and Space Administration, National Institute of Standards and Technology) should promote database and knowledge base construction and standardization, design-knowledge tool demonstrations, and pilot projects as part of their future systems programs. These programs should integrate existing computer-aided system tools. *Two potential ways in which this might be accomplished are to provide (1) funding for demonstration programs with creative problem solving and design concepts to teams of university faculty and students composed of computer scientists, engineering design specialists, materials scientists, and cognitive psychologists and (2) financial incentives to industry for sharing materials property data where input to public and limited access materials knowledge bases can be controlled.*
- *Industries and universities* should be encouraged to collaborate in:

EXECUTIVE SUMMARY

1. developing and using well-defined standards for electronic information sharing to enable selective protection of organizational private data, company proprietary data, and industry restricted data from the public domain data;
2. improving contact between researcher, designer, and supplier on design teams;
3. increasing rate of generation, validation, and exchange of materials data;
4. developing powerful programs for service life prediction of structural components from materials data, constitutive models, and in-service nondestructive testing;
5. developing models of practical significance to product development;
6. providing materials development data in machine readable electronic format;
7. preparing standards for knowledge representation of materials information (e.g., properties tables, graphs, and pictorial descriptions of microstructures);
8. publicizing success stories where experienced engineers select materials showing that proper representations together with reasoning examples will promote effective material computer-aided systems development; and
9. developing an information base on available (network accessible) materials databases and computer-aided systems.

EXECUTIVE SUMMARY

1

Introduction

The selection of the proper materials for a structural component is critical to engineering design. Existing design procedures may currently be sufficient, especially where experience exists, but fierce industrial competition is spurring the search for improved methods and tools. The main drivers are quality, life-cycle cost, and time-to-market. Improved design efficiency and accuracy may have an enormous impact on the economic viability of the final product.

Materials selection is governed by many factors, some of which are often in opposition. The principal selection factors include the service requirements and design life of the product; the availability of candidate materials and the appropriate data on application-specific properties for them; the company's *make* or *buy* decision for the system components; the customer preferences; and most importantly, the total life-cycle cost.

BENEFITS

Many designs initially fail because of a lack of relevant experience or because the design team did not include appropriate experts who "could have told us so." In the end, the rework associated with design requalification significantly increases cost and time-to-market. Thus, the use of computer-aided systems that assist design teams could potentially reduce product cost and time-to-market. Computer-aided systems for materials selection could assist concurrent engineering activities by helping to resolve many of the materials dilemmas presented during the initial design phase and by helping to guide the selection process based on the data and experience compiled from previous product development. Advanced computer technologies would also make it feasible to archive design experience as cases in a corporate knowledge base for subsequent re-use, tailoring, and evolution.

The development of a computer-aided system to support materials selection could also accelerate the general acceptance of new materials and processing technologies. The quality and efficiency of the materials selection process would be enhanced by increasing access to knowledge of factors such as materials options and life-cycle costs. For instance, designers could be provided with a range of possible materials and manufacturing methods for a proposed part, based on a given set of characteristics and cost-performance criteria. Thus, the members of the design team would not be totally reliant on their own personal experience and limited design-handbook information during the materials selection process but would have access to information on promising new materials and processing technologies that could be exploited.

Computing technology is no longer a barrier to the development of computer-aided systems for materials selection. Advances in reduced instruction set computer (RISC) chip technology already allow high-performance, inexpensive workstations to perform design layout, structural analysis, and materials processing simulation. It was generally believed in the early 1980s that the use of advanced modeling techniques, such as three-dimensional modeling, was not practical because of the large amount of computer time required for analytical simulations. Since then, computer speeds have dramatically increased. Accurate modeling and simulation of a unit process is currently becoming the norm. A range of new computer products are now available that enable the development of computer-aided systems for materials selection:

- high-performance microcomputers;
- high-performance workstations (minicomputers);
- workstation clusters;
- RISC parallel systems (e.g., 16 computer processing units);
- mainframe-workstation networks;
- vector supercomputing; and
- massively parallel computing.

The wide range of hardware capabilities will soon bring the cost of implementing computer-aided system logic and process simulation within affordable limits. The continued evolution of cost-effective, high-performance computing in conjunction with a national information superhighway infrastructure will further assist the nation's manufacturing sector in becoming more competitive in the international marketplace.

DEFINITIONS

Computer-aided systems are broadly interpreted in this report as advanced computing technologies that access various modules to provide specific information when requested by user input. A computer-aided system has three primary elements: (1) an interface with the user, (2) a reasoning element that triggers system action, and (3) a knowledge element in the form of databases, knowledge bases, and modeling modules that provide the information and analyses to be applied. Computer-aided systems for materials selection in design will contain only a subset of the total product knowledge applied during design.

Knowledge representation is the form in which facts and relationships are encoded and stored in the knowledge component of a computer-aided system. Knowledge representation serves five distinct roles. First, knowledge representation is a *surrogate* for knowledge. Second, it serves a set of *ontological commitments* or terms with which a computer-aided system can reason. Third, it is a *partial theory* of intelligence that expresses the fundamental concept of intelligence, the inferences that are possible, and the inferences that are made. Fourth, it is a *medium for pragmatic computations*. Fifth, it is a *medium for human expression* (Davis et al., 1993).

Concurrent engineering is "a systematic approach to the integrated, concurrent design of products and their related processes, including manufacture and support, [that] is intended to cause the developers, from the outset, to consider all elements of the product life-cycle from conception through disposal, including quality, cost, schedule, and user requirements" (Winner et al., 1988). The process of conducting design tradeoffs can be done sequentially or in parallel. In this report, the committee focuses on computer-aided systems that can support a design process in which design decisions are made in parallel, or concurrently, by several members of a design team. When the design team contains members that have access to all knowledge pertinent to the creation of the product, its use, and its ultimate retirement, that team is called an "integrated product-development team" (IPDT).

To practice concurrent engineering effectively, all knowledge related to the manufacture of a component and its maintenance in a delivered system must be available to the IPDT. Life-cycle data and experience knowledge is thus an important prerequisite for the full application of computer-aided systems in which design choices are evaluated. Because of their importance, the establishment of life-cycle databases is now required by the major Department of Defense initiative on Computer-Aided Acquisition and Logistics Support.[1]

STUDY OBJECTIVES AND SCOPE

This study concentrates on the materials-specific knowledge elements of a computer-aided system. The Committee on the Application of Expert Systems to Materials Selection during Structural design determined that the development of generic computer-aided systems is already receiving a great deal of attention within the computer science community. The basic information requirements for a computer-aided system for materials selection are receiving little attention within the materials community, however. Thus, the committee assessed that this study would have the greatest impact if it (1) detailed the capabilities required for computer-aided systems to be of value to the materials selection process during concurrent engineering, (2) identified the issues inhibiting the development of such a system, and (3) recommended materials-specific applications and developments in database, knowledge base, and materials modeling that would aid the production of a knowledge element appropriate for computer-aided systems for materials selection during design.

During the study, the committee examined engineering-related design decisions involving geometry or spatial relationships to determine how design rules incorporated materials data and how assessments of performance,

---

[1] Computer-Aided Acquisition and Logistics Support is a three-phase program that requires: (1) the adherence by contractors to data exchange standards; (2) the linking of contractor and government agency systems databases with strong emphasis on demonstration of concurrent engineering during system design; and (3) the development and automation of design knowledge bases (DOD, 1986).

processing, manufacture, and support (e.g., reliability and maintainability) were made. The state-of-the-art computer-aided systems that assist the materials optimization or design tradeoff processes were reviewed. The study also considered the use of materials modeling since it can potentially reduce the quantity of data required by the system. The committee used case studies to provide instances of design decisions involving geometry or spatial relations; design rules associated with performance, processing, manufacture, and supportability; and advanced computer technologies and concepts that aid in the optimization or design tradeoff process. The committee recommended the research and development (R&D) that is required in materials-specific databases, knowledge bases, and modeling to facilitate concurrent engineering design.

The committee composed this report for the study's sponsors: the Department of Defense and National Aeronautics and Space Administration. However, the committee also recognized that the report's audience included:

- structural engineers, materials technologists, and computer technologists in the sponsoring government R&D agencies who fund research, reduce technological barriers to agency projects and missions, and enable the transition of technology into products and processes;
- structural, materials, and computer scientists and engineers engaged in university and industrial R&D who strive to innovate in order to overcome technological barriers, demonstrate technological advancements, and enable the transition of technology into products and processes; and
- product design teams who aspire to maximize the quality of the design process and the resultant value of the product to the ultimate customer: the product's user.

2

Materials Selection in Structural Design

This chapter discusses the structural design context for materials selection, the materials selection process, the evolution of computer systems that support the design process, and the needs for materials information. The committee chose to describe the structural design process using aerospace vehicle design as a case study. This chapter is based, in part, on information gathered by the committee during a two-day site visit to the Boeing Commercial Airplane Group[1] and comparative data on the Grumman design process. However, the design process and information needs detailed here are generic and applicable to structural designs in many industries (e.g., buildings, bridges, oil rigs, automobiles, ships, and spacecraft).

CONCURRENT ENGINEERING AND DESIGN ORGANIZATION

The design and development of a structure like an aircraft is enormously complex. The original sequential approach to aircraft design was to break the structure and systems into manageable sections. Preliminary designs of each section were then evaluated sequentially and modified by a multitude of different engineering, manufacturing, quality-assurance, and operations-support experts (Figure 2-1). This sequential approach led to extensive changes and errors during and following the design process, problems with communications between the different disciplines, increases in development costs, and extensions in design and manufacturing schedules. Consequently, the amount of needed rework and redesign accounted for a significant proportion of production costs.

The concurrent engineering approach, supported by centralized digital databases for geometry, materials, fabrication, and assembly processes and paperless drawings, was proposed to improve the design process and reduce rework and redesign (Winner et al., 1988). Figure 2-2 compares the sequential and concurrent engineering approaches.

Boeing has implemented concurrent engineering through an approach that uses design build teams (DBTs). The DBT approach establishes an IPDT for designing new products and systems and executing a concurrent engineering/manufacturing process. The team goals are to produce error-free designs that are optimized in terms of performance, weight, and production and operating costs.

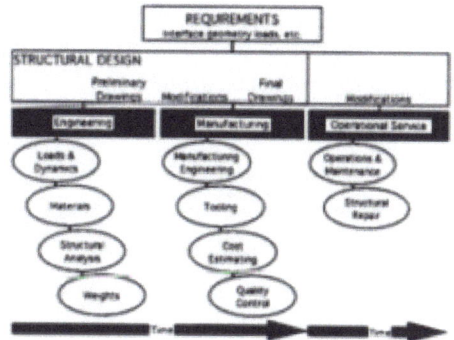

Figure 2-1 The sequential engineering approach to structural design. Reprinted courtesy of RNH Associates, Incorporated.

---

[1] The Boeing site visit comprised testimony from materials and structures experts, design managers, and automated systems specialists who were improving the design process via organizational and technological innovations. (Note: The selection of Boeing as a case study should not be interpreted as a statement that their design process is superior to those of other companies.)

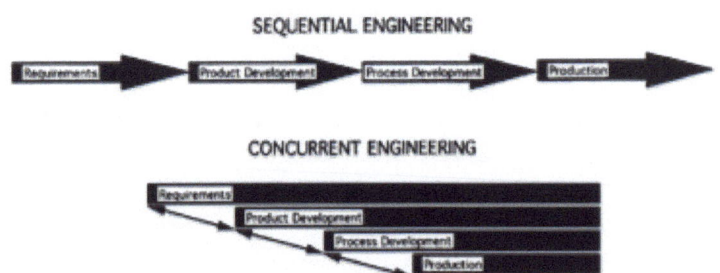

Figure 2-2 A typical comparison of sequential and concurrent engineering. Variations of this illustration are presented in Winner et al. (1988), Whitney et al. (1988), and NRC (1991).

The first step in the DBT approach is to divide the systems into major categories (e.g., structure, avionics, flight controls, mechanical systems, environmental systems, hydraulics, flight deck, and payload) as well as generic integration areas (e.g., airworthiness, reliability, and maintainability). These categories are then further subdivided. For instance, the basis structure divisions are body, wing, empennage, and propulsion system. These, in turn, are subdivided even further into manageable components and subcomponents, each of which is the responsibility of a separate DBT. For example, the main body components are cockpit, forward section, center section, rear section, and tail fuselage, as well as doors, door cutouts, floors, and floor beams. A typical hierarchical relationship between the IPDT and the DBTs is shown in Figure 2-3.

Boeing initially implemented the DBT system in an attempt to remain competitive in the global marketplace (NRC, 1993). The Boeing 777 program peaked at a total of approximately 250 DBTs, including 97 DBTs related to structures.

The structural DBTs (Figure 2-4) are composed of design, structures, materials, manufacturing (e.g., tooling and machining), quality control, and cost analysis experts; some teams also include liaison representatives from key subcontractors. Additional support may also be provided as required by representatives from other company divisions or by specialists on a part-time basis. The team members from the various disciplines responsible for creating a specific component or subsystem from conception through final design are collocated, and each team member is expected to participate fully in the DBT decision making process (Boeing Commercial Airplane Group, 1991). After the total design (including tool design) is completed, manufacturing is empowered to review and approve engineering data sheets verifying producibility prior to drawing release. A simplistic representation of the interactions within a typical DBT is shown in Figure 2-5. An important aspect of the process indicated in the figure is that optimization decisions are made from the perspective of the entire system, not from that of a particular subsystem.

Although concurrent engineering has considerably reduced rework, structural design and material selection remain iterative, cyclical processes. Structural analyses are performed on candidate preliminary design, and modifications are made to satisfy structural requirements. Weight and cost estimates are used for tradeoff studies to

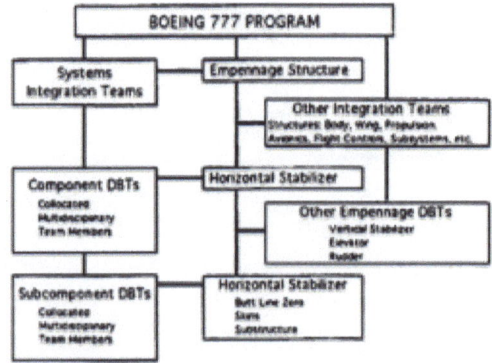

Figure 2-3 An example of a typical structures DBT hierarchy: Boeing 777 Horizontal Stabilizer DBTs. Source: Boeing Commercial Airplane Group.

Figure 2-4 The expertise in a structural design concurrent engineering team. Reprinted courtesy of RNH Associates, Incorporated.

identify and select the best materials and design configuration. The DBT approach also addresses most of the inter-disciplinary problems associated with candidate design concepts and material selection early in the preliminary design phase. Compromise solutions can then be identified and selected by the DBT members before the complete design is finalized.

Each DBT records team notes, memoranda, and summaries of project reviews in DBT libraries. These can be accessed by other DBTs to obtain information and digital design data. This allows rapid dissemination of changes that affect the interface between components, facilitates tradeoffs using global criteria, and ensures storage of lessons-learned data for future designs. These records are primarily found in hard-copy form. Although some are filed electronically, they are not available for on-line reference.

MATERIALS SELECTION DURING COMPUTER-AIDED DESIGN

A DBT team requires an enormous amount of detailed information to develop structures that will satisfy performance, reliability, safety, weight, and durability requirements at economical production, operation, and maintenance costs. In the 1960s, structural design and analysis consisted of slide-rule and adding-machine calculations using formulae and tables from handbooks in combination with numerous assumptions based on prior experience. The resulting designs were then evaluated by materials, manufacturing, and cost-estimating personnel who fed back their recommendations for design changes. Design and engineering operations are currently performed rapidly and accurately by DBT members using interactive computer-aided engineering or computer-aided design and computer-aided manufacturing (CAD/CAM) programs. Many different integrated computer-aided engineering and CAD/CAM systems are currently available. Even the most advanced of these focus only on finite element analysis (FEA) or boundary element analysis computer programs and currently have little materials selection expert system capabilities.

Computer-aided engineering and CAD/CAM systems generally use a mixture of bars, panels, and solids, which are utilized from preliminary design through drawing release. The structure is predominantly modeled using combinations of bars and panels for the structural analysis and optimization programs because of the significantly longer computer times needed to model the structure and complete the analysis or optimization using solids. Solid elements are only used when the structure cannot be

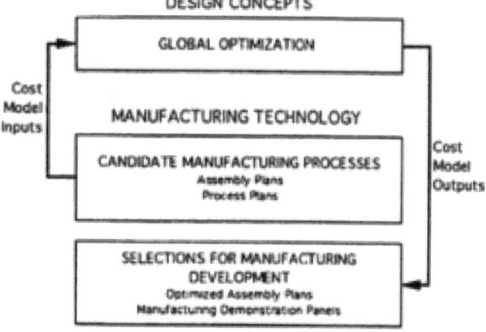

Figure 2-5 The interactions of a typical DBT during initial concept development. Source: Boeing Commercial Airplane Group.

realistically modeled using the simpler elements or when more accurate determination of the three-dimensional state of stress or strain in the component is needed. For example, Grumman used the model of a wing for the X-29 and associated FEA (Figure 2-6) in combination with a fuselage model to determine the loads in the structure and the dynamic and aeroelastic behavior of the wing required to preclude divergence and flutter. Aerodynamically induced structural divergence was avoided by designing the carbon- epoxy covers to provide bending-twisting coupling to the wing, taking full advantage of the anisotropic properties of the composite material. This model was iteratively appraised by structural analysis, weight optimization, and divergence analysis computer programs to determine the geometry and orientation of the carbon-epoxy tape for each of the 148 plies in the upper wing skin and the 158 plies in the lower wing skin. The same model and computer programs were then used for selection of the materials and the sizing of the cap areas and web thicknesses for the other wing components. As shown in Figure 2-6, the wing covers are carbon- epoxy. The other materials used in the wing component are steel, 6A1-4V titanium, 2024 aluminum, an woven glass-epoxy (Hadcock, 1985).

Three-dimensional models of forgings or machined parts are used for more detailed analysis and sizing of components, such as complex wing-to-fuselage attachment fittings and control surface hinges. These models predict the boundary loads and constraints from the overall FEA. The information from these programs can be electronically transferred to CAD/CAM systems to generate the drawings of the detail parts and assemblies for manufacturing engineering.

In all these programs, material properties and external geometry are generally input data. Structural optimization is done iteratively. Structural geometry, which depends on material properties, panel thicknesses, and stiffener sizes,

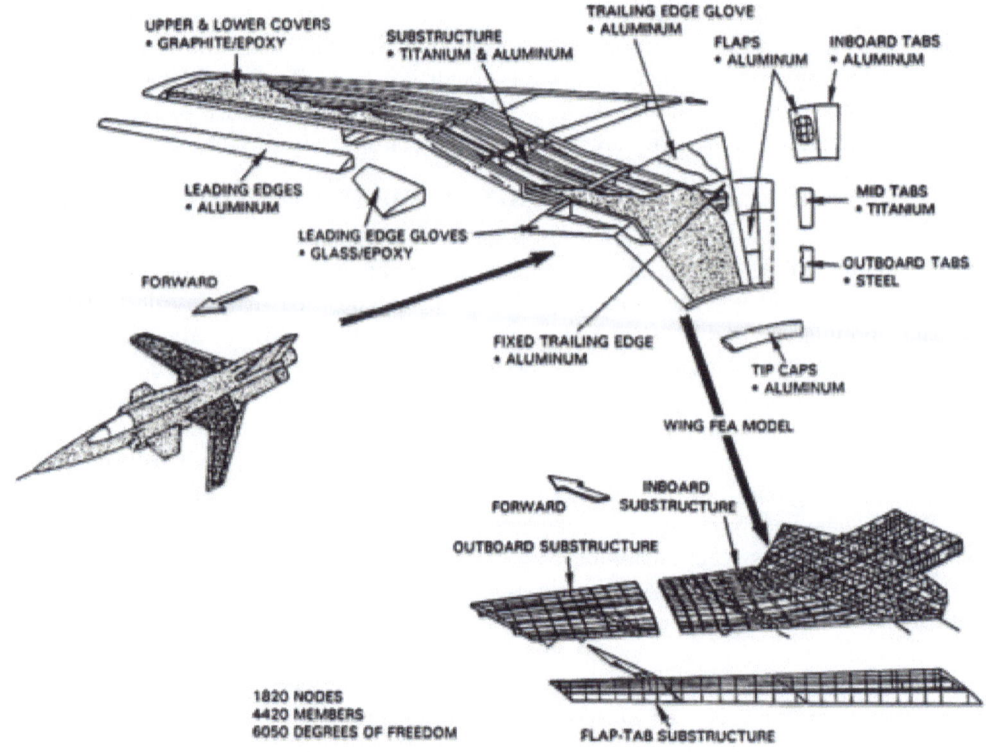

Figure 2-6 A model of the wing of the Grumman X-29 and associated FEA. Source: Northrop Grumman.

can be automatically adjusted during the iterations. The effects of changes in materials selection can be evaluated by executing the programs with different materials-properties data sets. This design tradeoff analysis process can be very time consuming, particularly when there are large numbers of candidate materials for each part and a range of structural analysis tests, such as thermal strains; dynamic behavior; fatigue; fracture; durability; and, in the case of combat aircraft, survivability. However, optimization programs are emerging that will allow the selection of *best* choices given the constraint parameters specified by the design engineer.

The aerospace industry has traditionally adopted a rigorous, yet conservative, materials selection process to minimize the risk associated with the introduction of new, and therefore less-proven materials. Risk as a factor in materials selection will be discussed in Chapter 3.

Some integrated computer programs are available for design, structural analysis, and production of complex-shaped castings and injection-molded plastic parts. These programs include thermal and flow analysis of the liquid material, design of patterns and molds that may include cooling passages to eliminate distortion and cracking during cooling, and determination of residual strains (see Appendix B).

SUMMARY OF MATERIALS INFORMATION REQUIREMENTS IN DESIGN

Table 2-1 provides a partial listing of materials-related information that is needed in the materials selection process. Materials selection is strongly influenced by overall product design, manufacturing, and cost requirements. Some of the product design requirements for aircraft structural design are presented in Table 2-2.

The major structures and materials design interactions are shown diagrammatically in Figure 2-7. Material selection is directly or indirectly defined by the combination of these design interactions. These interactions include most of the information needs of a team to design and select the materials for a primary structure component.

A summary of designer *wants* pertinent to the application of expert systems in the materials selection process during structural design is listed in Table 2-3. This table provides the basis for establishing the range of information technologies pertinent to the materials selection process that will be assessed in the next two chapters.

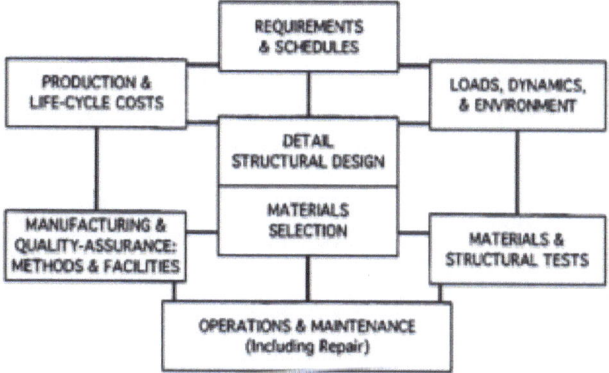

Figure 2.7 Structures and materials interactions. Reprinted courtesy of RNH Associates, Incorporated.

Table 2-1 Examples of Materials Information Required During Product Design

Material identification	Temperature (cryogenic–elevated)	Joining technology applicable
Material class (metal, plastic, ceramic composite)	Tensile strength, yield strength	Fusion
Material subclass	Creep rates, rupture life at elevated temperatures	Adhesive bonding
Material industry designation	Relaxation at elevated temperatures	Fasteners
Material product form	Toughness	Welding parameters
Material condition designation (temper, heat treatment, etc.)	Damage tolerance (if applicable)	Finishing technology applicable
Material specification	Fracture toughness (define test)	Impregnation
Material alternative names	Fatigue crack growth rates (define environment, and load)	Painting
Material component designations (composite/assembly)	Temperature effects	Stability of color
Material production history	Environmental stability	Application history/experience
Manufacturability strengths and limitations	Compatibility data	Successful uses
Material composition(s)	General corrosion resistance	Unsuccessful uses
Material condition (fabrication)	Stress corrosion cracking resistance	Applications to be avoided
Material assembly technology	Environmental stability	Failure analysis reports
Constitutive equations relating to properties	Toxicity (at all stages of production and operation)	Maximum life service
Material properties & test procedures	Recyclability/disposal	Availability
Density	Material design properties	Multisource? Vendors?
Specific heat	Tension	Sizes
Coefficient of thermal expansion	Compression	Forms
Thermal conductivity	Shear	Cost/cost factors
Tensile strength	Bearing	Raw material
Yield strength	Controlled strain fatigue life	Finished product or require added processing
Elongation	Processability information	Special finishing/protection
Reduction of area	Finishing characteristics	Special tooling/tooling costs
Moduli of elasticity	Weldability/joining technologies	Quality control/assurance issues
Stress strain curve or equation	Suitability for forging, extrusion, and rolling	Inspectability
Hardness	Formability (finished product)	Repair
Fatigue strength (define test methods, load, and environment)	Castability	Repeatability
	Repairability	
	Flammability	

Table 2-2 Typical Product Design Requirements for Aircraft Structure Development

Performance	Cost	Testing
Design loads and conditions	Design	Load-temperature-environment spectra fatigue
Associated air loads and accelerations	Production	Quality
Fuel usage	Preparations	Repair and reinspection
Cabin and cargo hold loadings	Material handling	Automated and nonautomated quality-control equipment
Temperatures and associated environmental data	Safety	Vendor/supplier qualification for new materials part fabrication
Fatigue spectra	Environmental and waste disposal	
Fail-safe and safe-life design	Interfaces	
Aeroelasticity requirements	Geometrical tolerances	
Airworthiness standards and design requirements	Structural assembly	
(Federal Aviation Administration: Federal Aviation Regulations, Advisory Circulars, etc.)	Surface smoothness and tolerances	
	Avionics	
	Propulsion	
	Environmental control	
	Passenger accommodations	

Table 2-3 Summary of Designer Wants

Design Tools	Design Cycle Time/Time to Market	Expert Agents
Material/processing/manufacturing tradeoffs in concept design	More tradeoffs considered in given time Iteration for realistic materials targets	Gather pertinent design information from multiple sources
Composite materials structures design tools	Reduce cycle time to market	Specific expert systems for each component design team
Quality materials-selection aids	Rapid deployment of new material	
Design Knowledge	Risk Reduction	
Information on the competition	Trusted design and materials data	
Lessons-learned knowledge base	Reduced risk in selecting new materials/processes	
Materials-use case base indexed by multiple attributes	Production Capabilities	
Cost Knowledge	Facility availability	
Cost models	Equipment availability	
Life-cycle costs	Workforce experience capability and availability	
Manufacturing costs	Viable supplier options	
Material prices		

3

Enhancing the Materials Selection Process in Design: A Vision

To identify the information technologies required for a computer-aided system to support materials selection, the committee articulated a future vision of a full-function Computer-Aided Materials Selection System (CAMSS) based on the information summarized in Chapter 2. In the future, materials selection is envisioned in a business context that has several major differences compared to current environments. The engineering design process is evolving from a stage of emphasis on concurrent engineering (i.e., the simultaneous design of products and manufacturing processes within a company) to one on concurrent enterprise processing (i.e., the simultaneous design of products and processes that takes into account internal and external partnering, preferred supplier relationships, and corporate alliances). There is an ever-increasing pressure for accuracy, flexibility, speed, and competitive leadership. The information systems supporting the concurrent enterprise process will be more ubiquitous, powerful, and integrated into the business process.

As a result of these changes, the impact on materials selection is substantial. Materials selection will be based on a much broader range of concerns and not on isolated, sub-optimized steps. The concurrent enterprise process demands that material selection is not only broad based, but done fast, right, at the correct time, and once. Software to support materials selection will be part of an integrated computing environment that spans the concurrent enterprise process and makes use of embedded assistance for many aspects of the product life-cycle.

Amidst the evolution of the business context, materials selection continues to occur in two forms: strategic materials selection and routine materials selection. There are no sharp distinctions between these, but strategic decisions are primarily in response to corporate objectives, high-visibility customer requirements, or strategic technology planning. The introduction of new materials or processes for a particular product application is nearly always a strategic decision. Such decisions are strategic because of the time required and capital costs associated with validation and investment in new process capabilities.

The pressures for increased agility in response to global competition is a mixed blessing for the introduction of new materials. The competitive pressure places demands on leadership but strips away the time to react. The advanced computing support for strategic and routine materials selection differs but shares a common infrastructure. The following four sections explain the vision for this common business and information processing environment, discuss the unique capabilities required for strategic and routine materials selection, and examine the basis for innovative materials selection in design.

INTEGRATED ENGINEERING SUPPORT IN INTEGRATED ENTERPRISES

Enterprises consideration are causing organizations to change the way they view themselves. New relationships with internal and external units are emerging. The shift to an emphasis on concurrent engineering is evidence of the shift that concentrates on internal and external partnering. External partnering has led to favoring preferred supplier relationships over low-bid competition. Technology "food-chains" are being addressed with corporate strategies for strategic technology planning. Alliances with technology suppliers are increasing.

As more cooperation emerges between units, more unified communication and computing environments are ramping up to meet the need. Increased emphasis on standards gives evidence to this shift. For example, the Initial Graphics Exchange Standard, which is used to exchange geometry, is expected to be supplanted by the evolving, international Standard for the Exchange of Product Definition Data, the goal of which is the exchange of complete, unambiguous computer-interpretable definitions of the physical and functional characteristics of a product throughout its life-cycle. These shifts provide an infrastructure to support future concurrent enterprise processes. In the mean time, computing hardware and software capabilities continue to evolve with an emphasis on integrated computing environments and open systems. Partnerships between software suppliers permit engineering organizations to consider a suite of applications that collectively cover substantial acreage in the art-to-part landscape. Standards in user interface technology (e.g., X-windows) are breaking down conceptual barriers between computer applications. In a few places currently, and more so in the future, the engineer is supported by a computing environment for the rapid transmission of shared data that links to other engineering and manufacturing organizations both within the company and with suppliers and vendors.

In the future, functional capabilities of software will be integrated so that conceptual design alternatives can be developed and evaluated for any number of criteria during early product planning phases. The alternative selected in this process must have a high probability of being manufacturable at the target costs negotiated by the product team. There must be an equally high probability that the product meets the requirements of the customer and is aligned with the technology plans of the enterprise.

The development and evaluation of these alternatives by design teams could be assisted by integrated computer-aided systems. These knowledge base systems are woven into the computing framework. Because of the integration of the environment, they do not appear to users as separate systems but rather add to the functionality that the system provides. Thus, from the user perspective, knowledge base tools are undifferentiated from analysis and design automation tools.

Typically, computer-aided systems are to provide design advice, leaving final decisions to the engineer. The advice given by the system can be as simple as selecting a default material specification. It can tell the engineer where to find suppliers of relevant material or it can retrieve that material. It can analyze a design and provide a quantitative or qualitative judgment. It can suggest an improvement or generate an alternative that includes the improvement. It can search a variety of alternatives and suggest the best or the best few alternatives based on quantitative or qualitative judgments and user-supplied criteria. The alternative selected during concept evaluation is then available for refinement in a coarse-to-fine development process. The details of the computer model of the product then evolve, aided by the use of a number of design advisor tools that provide reference information, analyze a design, critique it, improve it, or optimize it for a given set of design metrics. Analysis and design automation tools also help in the refinement of the product description.

Available throughout the process are advice and knowledge about materials. Significant material properties as well as emerging considerations, such as life-cycle costs, are to be increasingly supported by the information infrastructure. Materials knowledge is to be accessible to the engineer as reference material through design advisors that interact with the user as well as product and process models to analyze, critique, improve, or optimize it. Major tools in the integrated environment should provide the following capabilities: managing electronic repositories of data and documents; searching through past development histories to find similar or analogous products; managing interactions with other parts of the enterprise; managing requirements; predicting performance characteristics; predicting manufacturing and maintenance characteristics; estimating costs; suggesting improvements to the proposed product or process description; and releasing material, product, and process descriptions to other components of the enterprise.

SUPPORTING STRATEGIC MATERIAL DECISIONS

As indicated earlier, strategic materials selection almost always occurs for new material introduction. It also occurs when there are several material alternatives that represent significant tradeoffs in critical customer requirements (e.g., appearance, durability, cost, and risk).

In the future, materials experts should use simulations of material performance at both micro- and macro-structural levels to reduce the cost of material validation. Material models must include manufacturing process performance as well as product performance. Major cost savings are found in the reduction of decision-time and rework required. Material supplier and users must regularly join together to develop and specify materials and processes.

Strategic decisions are to be made using formal decision methodologies and computer tools for support. Quality Functional Deployment and Decision and Risk Analysis are two examples of methods supported by tools (see Chapter 4). Materials are the focus of many strategic decisions but only one of many factors in far more strategic decisions. To support the decision process, performance and cost models for materials and processing are to be used for strategic decisions that include long-range business planning.

Strategic decision making cannot be handed over to computers; rather computers and information systems must be relied on to provide access to documented information and models of the business, product, processing capabilities, and processing influences on materials. They can also manage the complexity of related decision variables and keep track of alternatives that are under consideration. Broader access to such information contributes to a better understanding and quantification of the risk from the introduction of new materials.

It is important to recognize that materials design for "structure critical" applications tends to be rather conservative. Designers cannot afford to take unnecessary risks with new materials but they can gain expertise with processing and performance of new materials in noncritical or development applications (e.g., composite fishing rods, nitanol eyeglass frames, or ceramic scissors). Such experience is vital in gathering data and confidence for critical purpose applications. However, it is imperative that future systems be able to collect, organize, and distribute such lessons-learned experience.

SUPPORTING ROUTINE MATERIAL DECISIONS

Routine material decisions happen every day for every component developed by the enterprise. There is increasing emphasis on the process for making such decisions to assure consistency, accuracy, and reliability. Computer systems can provide assistance to the engineering community to follow established processes but should not lock the user into a rigorous framework that strips the user of opportunities to exercise creativity.

There is pressure to include more factors in all decisions. Materials selection is influenced by factors such as manufacturing; assembly; service; and environmental impact of material production, use, and disposal or recycling. Computer-aided advisors can help manage the complexity of the many concerns. Supporting the product team in the materials selection process are electronic documents; cost estimation tools; trade-study tools; material, product, and process data bases; and knowledge base systems that provide analysis, critiques, and product improvement suggestions. Materials selection falls within the scope of such tools.

SUPPORTING INNOVATIVE MATERIALS SELECTION IN DESIGN

A prospective computer-aided system should also be capable of assisting innovative design. It should not just provide a limited series of conventional material or processing choices. This section addresses the characteristics of CAD support systems for solving difficult, nonroutine design problems. The concepts presented here are drawn extensively from the recent publication by Steven Kim entitled *The Essence of Creativity: A Guide to Tackling Difficult Problems* (Kim, 1990).

A design problem can be characterized by its domain, difficulty, and size. Domain refers to the application area or areas, size refers to the amount of work needed to analyze and implement the design solution, and difficulty refers to the level of conceptual challenge to identify an acceptable solution. A difficult problem is one in which resolution is not readily discernible. Design problems can be ill-structured. They are generally not bounded by algorithmic models and may lack complete sets of heuristics to be applied to the design space. Therefore, an innovative design is the creative resolution of a difficult problem.

A creative solution exhibits certain features that are close in conceptual space and others that are more distant in that space—a concept Kim terms "the Multidistance Principle." Those aspects of the solution that are closest to the knowledge or experience of the design team may be clearly evident. Those aspects that are more distant in the solution space are the ones that often require insightful thinking. The Multidistance Principle has implications for the development of software tools such as computer-aided systems that enhance finding creative design solutions. These tools must be able to establish links to one or more attributes of the distant elements of the design solution as well as providing access to the more routine, detailed design features.

There are several factors that contribute to the generation of a creative design solution. The objective of the design must be defined and distilled into its elements in order to begin the design process. The elements of the design objective can be viewed as a logical hierarchy of design alternatives and decisions regarding the alternatives (Weber et al., 1991).

The creative solution process has two elements: (1) a structure, characterized by *diversity* and *relationships*, and (2) vehicles for enhancing the idea-generating process involving *imagery* and *externalization* . Diversity refers to the fusion of disparate ideas (i.e., the Multidistance Principle). It can be aided by memory enhancers, such as access to knowledge bases and historical archives. Relationships define the pattern among design space objects that can be discovered from reference to related problems and rapid enumeration of alternative solutions. Imagery is the generation of ideas through sensory images (auditory and tactile as well as visual). The most powerful imagery for humans is visual. It is possible through imagery to represent numerous objects and their relationships simultaneously. Externalization is the expression or communication of the ideas to others through text, models, and diagrams of the process. Externalization helps to clarify the idea and is often the most important step in achieving a creative design solution. The use of virtual reality is an example of the combination of imagery and externalization.

The creative solution process structure represents the important ingredients for innovative design and problem solving. The strategy for enhancing innovation is the development of tools that promote or enhance elements of this structure, especially memory, imagery, and externalization. With this strategy, one can identify several domains of knowledge that can contribute to the enhancement of innovative design:

- artificial intelligence (learning, inferencing, and knowledge representation);
- computer and network technologies (parallel processing, storage media, network communications, and workstations);
- human interface technologies (graphics, vision, touch, animation, simulation, and voice or speech); and
- cognitive psychology (perception, memory, reasoning, and insight).

Human memory, either in an individual or within a group, is both a store of archival knowledge and work area for the development and examination of design alternatives. The contents of human archival memory, enhanced by computer recall of details or related concepts, facilitates the generation of novel elements of a possible design solution. The working memory of the individual or group is a basis by which to craft the full solution. Imagery and externalization that aid in representing various solutions are key to bringing a wide range of information to bear on the design problem at hand. While computer-aided systems have been used to enhance logical, rule-based thinking and neural networks can *learn* perception, the element of cognitive psychology called "insight" is the key to the discovery of creative solutions to difficult design problems. Computer technologies and tools may not be able to replace human insight but could enhance it. This area needs research emphasis as a critical component of design technology.

SUMMARY

The materials-specific information technologies that designers require in a CAMSS and some of the computer technologies that are needed to build this system are summarized in Table 3-1. Figure 3-1 specifies the high-level conceptual architecture and some of the contents of a full-function CAMSS based on the vision presented in this chapter. The state of the art of the information technologies pertinent to the materials selection process is discussed in the next chapter.

Table 3-1 Summary of the Materials-Specific Information Technologies and Some of the Primary Computer Technologies Required for a CAMSS

MATERIALS SELECTION CAPABILITIES REQUIRED	PRIMARY COMPUTER TECHNOLOGIES REQUIRED
Routine Materials Selection—Standard selection process for every component developed that consistently, accurately, and reliably follows established procedures without eliminating opportunities to exercise creativity. System requires tools that manage the complexity of manufacture, assembly, inspection, service, and environmental impact considerations of material production, use, and disposal/recycling, and suggests product improvements.	Materials databases and knowledge bases Electronic documentation of previous designs Heuristics and selection reasoning traceback Cost estimation modeling
Strategic Materials Selection—Decisions to introduce new materials based on understanding/quantification of risk, impact on customer requirements, and correspondence with enterprise objectives and strategic technology planning. System should not make strategic decisions but should (1) provide design advice; (2) develop and evaluate conceptual design alternatives that meet customer requirements, manufacturing target costs, and enterprise strategic plans; (3) provide access to material-performance models and decision methodologies; and (4) collect, organize, and distribute lessons-learned experience to assist future decisions.	Modeling systems for tradeoff analysis: Constitutive Analysis Process modeling Performance simulation Cost estimation Risk assessment Lessons-learned collection and searching systems: Object-oriented databases Neural networks Electronic documentation Case-based reasoning
Integrated Enterprise Processes—Materials selection based on broad range of industrial competitiveness considerations, including internal and external partnering and preferred supplier relationships and alliances. Requires ubiquitous, powerful, and well-integrated systems with standardized computing environments and data structures to promote unified communication and permit rapid assembly/transmission of shared data both within companies and with suppliers/vendors.	Standardized data structures or data dictionaries Inter- and intra-company networking/communication systems
Innovative Materials Selection—Must assist innovative design and help solve difficult, nonroutine problems as opposed to just providing limited series of conventional material or processing choices. Tools that enhance/stimulate creative processes and insight via human interface technologies that promote learning, inferencing, and cognition.	Artificial intelligence (learning, inferencing, and knowledge representation) Database and knowledge base acquisition Human interface technologies (graphics, vision, touch, animation, simulation, and voice or speech) Linking of multiple knowledge bases

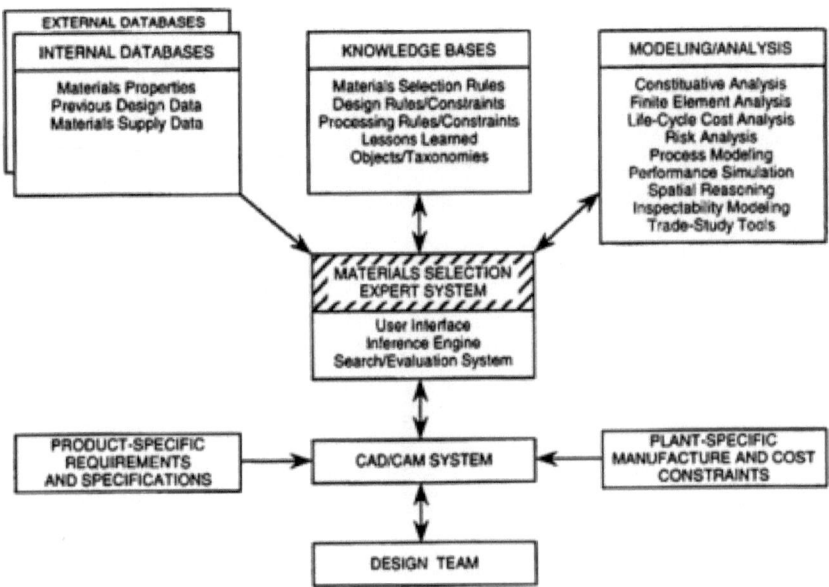

Figure 3-1 The conceptual architecture of a CAMSS.

4

Information Technologies Pertinent to the Materials Selection Process

This chapter discusses the key materials-specific information technologies required to produce the CAMSS diagramed in Figure 3-1. This chapter is divided into two sections: "Databases and Knowledge Bases," which pertains to the first two boxes at the top of Figure 3-1, and "Modeling and Analysis Systems," which pertains to the third box at the top of Figure 3-1. This chapter focuses on the use of computer-aided systems as tools or aides to design teams. Full-function automated systems currently require significant break-throughs in areas of frontier research and are particularly weak in tasks demanding creative insight.

DATABASES AND KNOWLEDGE BASES

As shown in Chapter 3, a CAMSS requires access to and application of materials databases and knowledge bases at every stage of use. In the ideal case, electronically stored knowledge about materials and design details could be provided automatically to the design team from databases and knowledge bases at appropriate levels of sophistication. This section provides a brief overview of the levels of knowledge representation in the automation of technical memory and discusses the principal methods for representing materials knowledge within a CAMSS to facilitate the design process. Appendix A contains a brief overview of some of the knowledge representation techniques discussed in this chapter.

Levels of Representation

The basic level of electronic knowledge representation is an electronic library (i.e., databases and knowledge bases). In this scenario, a human designer would perform essentially the same tasks as previously, but instead of searching for information through written material, the search would be conducted through screens. The screens could present prior designs, lessons learned, design guidelines, or standards. While electronic searches have advantages over manual searches, the cost of implementing all the necessary reference material electronically would be high and probably could not be justified based solely on productivity gains. However, since most documentation is currently being created electronically, this is a significant issue only for older reference material.

In its simplest conception, the electronic database or knowledge base would have no more embedded reasoning power than books (i.e., the user supplies all the reasoning). Three advantages of this basic type of knowledge representation are that (1) it is easy, in principle, to implement; (2) it is represented in natural language, with all its flexibility; and (3) it automatically makes the most recent versions of material available.

A higher level of sophistication would, continuing the analogy, consist of a reference book that opens automatically to the desired page and then, based on a user request, highlights that part of the page which the user needs. This requires a mechanism or process for the user to describe the reason and criteria for searching the reference book.

In a limited sense, such searches have been available for a long time in databases and knowledge bases using key words and indexing schemes. While keywords and index schemes are useful in restricting the material that will later have to be scanned by the human expert, these schemes are not very intelligent because a great deal of information is often presented that is totally irrelevant to the problem at hand. Keywording, moreover, works on alphanumeric information but is not easily adapted to other information types such as shapes, colors, and graphs. The technology to support multimedia, interactive reference books is now emerging (ACM, 1993; IEEE,

1993). This capability is an advance in that the author has pre-programmed expected search requests.

A higher level of sophistication of electronic representation of technical memory would be the equivalent of an educated assistant or technician that can search the library and retrieve the pertinent information in the background without direct user involvement. Representing the knowledge that an electronic assistant contains and representing the knowledge that the user requires are two different problems, however. For the materials selection problem in structural design, the intelligent electronic assistant would have to understand, at some level of competence, the information provided by the human experts. That information would consist not only of concepts used in materials science and engineering but concepts related to the entire life-cycle of design, manufacturing, inspection, and disposal or recycling. In the ultimate case, the electronic assistant will have to know the languages of many different pertinent databases and then be capable of representing that knowledge in a consistent form. This leads to the need for the intelligent integration of information from these multiple sources.

The highest level of sophistication envisioned would provide a full-function, computer-aided electronic assistant or technician who could not only find the correct reference material but also apply the results to the query to the design problem at hand. Just as we can imagine human assistants of different levels of skill, so we can also imagine electronic assistants at different levels of utility. As mentioned earlier, full automation of databases or knowledge bases to perform the complete design task is not currently feasible. However, certain select routines or computationally intensive tasks now performed by people can be performed by electronic assistants to provide particular advice or to critique selected aspects of the design. For instance, there has been considerable research into Agent Technology, whereby a user can specify an agent to roam the Internet to obtain appropriate information (ACM, 1994). This technology requires the use of data dictionaries or mediators, however, to recognize and translate the relevant information in different databases and knowledge bases and to integrate possibly conflicting data from multiple sources. This level of capability has been shown to greatly benefit the overall performance of the design teams in several limited instances (Klahr et al., 1987; Famili et al., 1992). For example, the three areas of expertise—product design, materials selection, and manufacturing—cannot be entirely separated for highly engineered products. Materials properties depend to an extent on the processing route, and processing considerations can be influenced by design constraints. Similarly, the design must reflect the reality of available material properties, and the properties are not completely independent of the design application (e.g., high loading rates can reduce a material's fracture toughness).

Issues Concerning Knowledge-Base Development

Reliance on standard databases that contain physical and mechanical properties is inadequate to support materials selection processes fully. Knowledge bases are required that capture the advantages and limitations of materials, their processability, and their application histories—all of which are critical to the design process. There are problems related to the definition, development, and construction of knowledge bases, however.

Definition of Knowledge Bases

While much has been written about knowledge bases, there is little agreement on the scope of what exactly constitutes one. There are wide variations in the literature describing designs and implementations of computer-aided systems.

Many companies have custom-built basic systems for their own applications. Table 4-1 summarizes representative computer-aided system application areas that relate to the materials information used in the design process in a manner consistent with the vision discussed in Chapter 3 . These applications are considered state of the art in the sense that examples can be found either in use or under development at major government and industrial sites.

Table 4-1 shows that there is a wide breadth of knowledge base applications. The list is also incomplete, since it only represents what is currently possible in the design and engineering phase of product development. If the scope were broadened to other phases of the product life-cycle, more applications could be listed that would require knowledge (e.g., diagnosis of the manufacturing process). Further work is required to determine what constitutes a knowledge base and how it differs from simple databases.

Table 4-1 Representative Applications Based on Knowledge of Materials

Application	Description	Knowledge Required
Selecting Materials	Assist the product designer in selecting the best material for the product.	Materials, performance, and processing knowledge.
Cost Estimation	Determine the manufactured cost of a product based on a parametric description of it. Could include full life-cycle costs.	Cost of raw material and costs associated with design, validation, manufacture, and disposal methods for various materials.
Manufacturing Process Planning	Develop a process or assembly plan for a product. The depth of detail can vary from process routing to instructions for controllers.	Manufacturing methods and processing data associated with various materials.
Predictive Analysis	Predict the performance of a product. Many dimensions of performance (e.g., stress, wear, and kinematics) can be analyzed.	Materials properties and characterization of usage of materials.
Manufacturability Advisor	Critique a product design, process plan, or operations plan regarding the efficiency, product quality, and cost of production.	Manufacturing knowledge about materials and the manufacturing processes associated with the materials.
Requirements Allocation and Balancing	Provide assistance to the product designer to allocate and balance requirements in formal requirements processes such as Quality Function Deployment.	Knowledge of the relationship of materials to the characteristics defined in the requirements flow down process. Examples include durability of materials, surface quality for finishing, and mass.
Search for Prior Designs/Products	Assisting product designer in finding most similar products from a library of products to take advantage of prior experience. Prior products using similar materials may be interesting even if product was quite dissimilar.	Any material properties may be of interest. Knowledge of similarities between materials and relationships between materials and product and process performance and cost. Repository of prior designs and product information.
Searching Standard Components	Assist in selecting a standard component from a library of standard components.	Any material properties of the standard components that are relevant to the cost and performance of the product.
Tolerance Analysis and Allocation	Predict the variability of the manufactured part with respect to geometric dimensions and other key product characteristics.	Knowledge about the relationships between materials and the manufacturing processes and manufacturing equipment.
Prototyping	Assistance can include selecting the best material for the prototype and the prototype tooling based on the materials used in the product. The prototype can be constructed with varying degrees of production-intent tooling.	Knowledge of any material properties and similarities of materials to be used for prototype and prototype tooling to those to be used in production.
Robust Design	A formal three-step process for producing high-quality, low-cost products.	Knowledge of material properties are used in the system design phase in deciding required features, functions, and product parameters. Knowledge of materials is used in the parameter design process to consider reliability and manufacturability of the components. Finally, in the tolerance phase, material properties are used to adjust product parameters to achieve broad ranges for manufacturability.
Trade-Study Methods	Assist the product designer in making choices with respect to product features, function, and manufacturing.	Knowledge of materials and how they relate to key product characteristics can be used in the evaluation of alternatives.

Development of Knowledge Bases

The absence of specific guidelines for the building of databases and knowledge bases about materials and the higher software development costs necessary to obtain generality and robust performance are barrier to the quick and effective proliferation of the use of databases and knowledge bases in material selection (or other fields as well). Knowledge base applications are currently developed in several different ways.

Some basic, commercial, knowledge base applications exist that the user community can acquire, install, and customize by loading site-specific models, information models, and information. One example of such an application is in the domain of cost estimation. This application allows the user to build parameterized product descriptions that incorporate relevant attributes that influence cost and manufacturability, material models that include parameters that impact cost, and process descriptions that link models of processes to the product and material attributes. The user can then employ a simple spreadsheet-like language to match processes and material to products and to compute cost estimates.

More advanced knowledge base systems can also be developed by using application shells or knowledge-engineering tools that implement knowledge base techniques developed by the artificial intelligence community to solve a variety of problems (e.g., diagnosis, simulation, and scheduling). An application shell is often a library of modules that can be used to assemble an application in some broad area and rapidly provide an environment for capturing and representing expertise in the form of rules and the knowledge structures. The advantage of using shells is that they permit the user to concentrate on representing the knowledge rather than attending to low-level programming tasks. Application shells do require tailoring, modification, and extension before use. Typically, a knowledge-engineering effort must first be undertaken to perform knowledge acquisition and routine programming tasks. For application shells, the cost of customization is often offset by the reduced cost for generality in the software. As with complete applications, many companies have internal products that they distribute to a number of sites.

Construction of Knowledge Bases

Two final barriers to the construction of knowledge bases are the higher hardware costs and the inherent nature of expert knowledge. The construction of knowledge bases require technologies beyond the standard manual entering of pertinent pieces of information. For instance, digital scanning combined with character-recognition technology can be used to enter application history, such as failure analysis results. Digitizing tables and figures using data capture software can allow each semantic element and semantic relation expressed in the tables and figures to be stored in a searchable file. Graphs can be made searchable for specific data, interpolated data, and extrapolated data. Audio-visual and multimedia digital storage and presentation are becoming common-place on engineering workstations. The appropriate assignment of recordings of design engineering discussions in knowledge bases will pose a significant research challenge. Annotations to diagrams relating key design decisions and constraints would also assist others in understanding the reasons for a particular choice.

Issues Concerning Database Development

The material databases needed during the design process are not openly available to U.S. industry. Some government and privately sponsored organizations, such as the Department of Defense's Information Analysis Centers and the National Materials Property Database Network (Appendix C), have made a start, but industrial data limitations have resulted in their falling short of the needs for knowledge base approaches to design.

The situation is currently little better than it was in 1983 when a National Materials Advisory Board report stated: "There is no national policy in the United States directed toward a rational system of materials properties data management and the situation in this area is best described as chaotic" (NRC, 1983). There is still no leadership for collecting, generating, validating, and updating the data needed for structural design. Most organizations will agree, however, that a certified database is a valuable asset. As recommended in a more recent National Materials Advisory Board study (NRC, 1993): "a federal agency, such as the National Institute of Standards and Technology, could establish a forum to develop the standards through timely, active participation by industry and other interested parties."

MODELING AND ANALYSIS SYSTEMS

As shown in Chapter 3, a CAMSS must have modeling and analysis systems to analyze the information available in databases and knowledge bases. This section examines the modeling and analysis systems pertinent to the materials selection process: Geometric Reasoning, Process Modeling, and Modeling for the Abstraction of Downstream Constraints.

Geometric Reasoning

Most CAD systems currently in industry are used primarily for drafting purposes with some analytical support. The level of graphic rendering has reached an impressive level, but all design decisions are essentially made by the design engineer with little or no decision support, or reasoning, other than visual feedback. Reasoning falls into the two major categories of synthesis (i.e., the systematic creation of alternatives as the product design and the design process become increasingly more specific) and abstraction (i.e., the elimination of possible design alternatives by downstream concerns and constraints, such as performance and manufacturability). The term abstraction is used since, in most instances, detailed downstream concerns (e.g., manufacturability, maintainability, or recyclability) need to be either simplified or abstracted to become more readily understandable by an engineer at an early stage of the design process. As discussed in the previous section, the designer should receive feedback about the consequences of a decision at a variety of different support levels and from a variety of different viewpoints (e.g., materials considerations, product performance, manufacturing, cost, service, and reliability).

Geometry plays an important role in design, yet many initial design decisions are made independently of geometric considerations. The final decisions in product design almost always involve form and geometric constraints. An example of such an information flow in an industrial setting is to imagine a car design team that is synthesizing a new car body. The current manufacturing approach would be to spot-weld several hundreds of pieces of sheet metal together to fit the part geometry. A possible alternative is an assembly of lightweight aluminum beams on which plastic panels are mounted. This design is called a space frame. However, to make an intelligent decision about whether this alternative design has merit, a design group must understand the essence of the aluminum casting and extrusion processes, the microstructures that result, the properties of the beams, and the methods for joining the pieces. This means simplifying the details of the process to the extent that someone with little materials expertise can take this information as the basis for performing design tradeoffs. The faster and more efficiently the manufacturing process and materials performance constraints are presented to the designers, the faster and more efficiently they can synthesize a new product.

When reasoning about materials and assigning material properties to certain geometric objects, several things are important. In principle, all geometric models, whether based on elements of one, two, or three dimensions, can be linked to attribute-value pairs, such as materials properties, like Young's Modulus, with a specific value assigned. Some of these attributes may be related to material properties or to the specifics of a certain microstructure. In the evaluation of the geometry, the attribute-value pair mechanism must be able to inherit property structures. Furthermore, when the attribute or its associated value assume a specific characteristic, the mechanism should be able to trigger events automatically. Most commercial CAD/CAM systems do not currently have this capability, although it is an active research area.[1] Significant research needs to be performed to create sound representational schemes with the described behavior, however.

Formal models are also not available at any level of representation that allows the derivation of structure-property relationships from first physics principles to an extent that a design engineer can benefit. One example is the field of dislocation theory (defect structure level). The flow stress (micro and macro material property level) of subgrain forming materials is understood to be proportional to the square root of the dislocation density, but the models currently available give only an order of magnitude relation for the proportionality factor in this relation.

---

[1] One such area is that of active database systems, which contain rules to monitor the state of the database and trigger operations to update the system or alert the user to certain situations (ACM, 1994).

A design engineer could benefit from such a materials science insight, however. It would help the designer to better understand the choice of a constitutive relation for FEA that is frequently used to determine the dimensions of load bearing components. The design engineer needs (computer-aided) decision-support tools that provide insight into materials science issues to consider material alternatives and processing tradeoffs effectively. Reasoning about materials needs to be closely coordinated with decisions regarding shape and geometry. Considering the earlier example of the car space frame, a novel design was created by combining material properties and processing knowledge with spatial layout.

Decision-support methodologies and tools that aid in synthesizing and finding constraints early on in the manufacturing process are also key to improving the quality and speed of product creation. Consider the following tool used for designing a side marker, a relatively simple component for a car (Figure 4-1). One side of this car side marker has strengthening ribs to increase the resilience. The system shown here recognizes certain geometric features that are important from a manufacturing materials viewpoint. Within that system, the designer receives feedback about whether the dimensions chosen are compatible with good manufacturing practice. Decision-support tools such as this can significantly reduce errors, cost, and development time. Their realization will largely depend on research performed to construct methodologies for representation in which geometric information can be properly linked with nongeometric information, like materials and processing knowledge and databases.

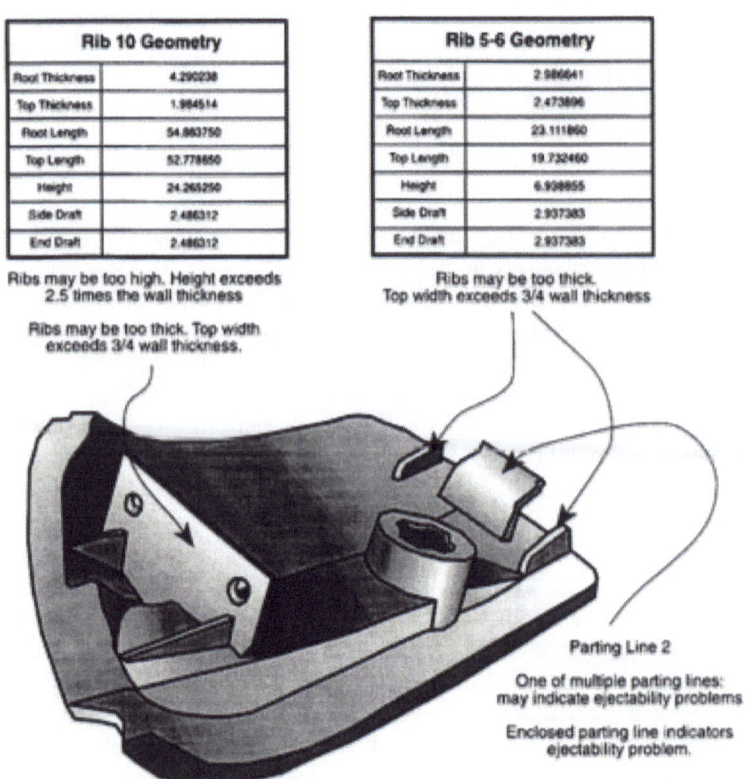

Figure 4-1 Automobile side marker.

Process Modeling

A process model is a mathematical representation or simulation of a process that allows problems to be solved in the computer rather than by empirical or experimental methods, especially trial-and-error techniques. *Simulation* is an important concept in the modern manufacturing organization. It uses mathematical models of real systems to test or predict the actual performance of the systems under various conditions. Through simulation, engineers and manufacturing personnel can test a design, analyze a procedure, or assess a process performance before implementing the real thing. Process modeling—discussed in the context of this report—deals with unit processes such as casting, forging, rolling, hot isostatic processing, heat treating, machining, chemical vapor deposition, and composite material fabrication. Constitutive modeling may be one of several elements in the overall process model. Constitutive models focus on predicting the mechanical response of a material as a function of prior processing history and internal structural parameters in response to externally applied forces. Process simulation could just as well apply to modeling the behavior of a Flexible Manufacturing Center or the simulation of the flow of information in a process plan. A simulation capability in the manufacturing setting can substantially decrease the energy, material waste, and time required to produce a product or implement a process.

A number of sequential steps are involved in any process modeling activity. These steps can be formalized and implemented in the computer as an activity model that is based on how the particular manufacturing enterprise does business, or they can be accomplished in a less formal mode of problem solving. The various steps involved in process modeling are listed in Table 4-2.

Process modeling is generally used to understand unit processes that require coupling of disparate physical phenomena. For instance, virtually all unit processes are governed by heat flow, fluid flow, plastic flow, stress, and phase transformations. These all can be modeled by a variety of numerical techniques. Several approximate numerical techniques for simulating material shaping and forming processes under arbitrary conditions have been used, including (1) the slab method (approximate stress analysis), (2) the slip-line method (method of characteristics), and (3) the upper-bound method (method utilizing an energy principle). Although these techniques contribute

Table 4-2 Steps in the Development of a Process Model

1.	Define the problem and state the problem-solving objective.
2.	Develop the mathematical model in accordance with the problem.
3.	Collect model input data and specifications.
4.	Implement process model in the computer.
5.	Establish that the desired accuracy or correspondence exists between the simulation and the real system.
6.	Establish boundary conditions for using the model.
7.	Run simulations to obtain output file.
8.	Post-process the output values to draw inferences and make recommendations to solve the defined problem.
9.	Implement and document the decisions resulting from the simulation and documenting the model and its use.

significantly toward understanding the mechanics of deformation in metalworking, they lack generality and often do not provide accurate estimates of the required forces and energy.

For fifteen years, the finite element method (FEM) has been applied to model a wide range of metalworking operations. FEM divides the volume of the plastically deforming material into a two-dimensional or three-dimensional network of discrete elements (finite elements). The deformation at selected points (nodes) is determined by the application of solid mechanics principles.

Specialty FEM analysis codes for process modeling currently have been developed for analyzing almost every class of unit-manufacturing process. The processes that have been simulated include machining, heat treating, sheet metal forming, shape rolling, ring rolling, extrusion, forging, powder consolidation and forging, superplastic forming/diffusion bonding, cogging, and radial forging. At least one of the commercially available three-dimensional codes is capable of concurrently modeling the equipment, the dies, and the workpiece's response to the boundary conditions, including the effects of different heat sources such as induction and resistance heating on the material flow behavior and the die reaction (Kiridena et al., 1989).

Almost every detail of a unit process can be modeled by FEM analysis, including predicting the evolution of microstructure and properties in the finished shape. The latter is made possible through the use of a technique known as dynamic material modeling (DMM), which uses constitutive relationships that define the evolution of metallurgical structure on a scale that ranges from the microscale of dislocations, precipitates, dislocation networks, and grains to the macroscale of laps, shear bands, and grain flow lines (Richmond, 1992).

DMM defines the intrinsic workability of the workpiece material at the macroscale level of structure in terms of mechanical and structural stability. This material model enables the process design engineer to define the control space for a stable process, including the number of preform shapes, the die velocity and temperature ranges, and the die geometries. Within the domains of the stable control space, where the activation energy is fairly constant, microscopic models can be used to predict structural evolution. The end result is a product having a controlled set of structures and properties in the finished shape.

As an example, the group of processes that can be described broadly as casting processes are now being simulated by several numerical methods. Processes of this type are investment casting, permanent mold casting, die casting, squeeze casting, and plastic injection molding. The approach to designing these processes by process simulation is fundamentally the same as for any other process.

A choice of numerical methods are available for this class of problem. The finite difference method, boundary element method, and FEM have been used for modeling the behavior of casting processes that are controlled by coupled thermal, fluid, and stress phenomena. Both the finite difference method and the boundary element method can be used when the material properties are linear and the product geometry is relatively simple. However, when the problem couples thermal, fluid, and stress phenomena, the FEM modeling technique is superior.

One example of incorporating material behavior during processing conditions is the use of experimental castability maps that are expressed in terms of the fundamental variables predictable by the process model. These maps provide a framework for modeling the casting process at the macro or continuum level of analysis. Like DMM, which predicts stable plastic flow, the castability maps define the domains where certain microstructures or defects will form when certain ratios of R/G occur, where R is the interfacial velocity and G is the temperature gradient at the solid/liquid front. These modeling parameters were derived from first principle understanding of the nucleation growth kinetics.

Micromodels can be integrated with the macromodels for predicting the evolution or occurrence of microstructural features. More effort is required to expand the number of micromodels to cover all possibilities, however. Mechanical property predictions are not yet possible because very few correlations have been made with microstructures and thermomechanical histories. These models suffer from not incorporating the knowledge of the basic physical mechanisms involved.

Models can be used to better inform suppliers on their process requirements. For example, cast products are usually considered to have inferior properties to wrought products because of the large variation in mechanical property values that can be found in the same cast product produced by different vendors. Process modeling can reduce the scatter from vendor to vendor by specifying to each vendor the desired thermomechanical history for a given component.

Thermophysical property data play a key role in modeling many unit-processes such as investment casting, welding, crystal growth, glass making, microwave processing, and composites production. A critical need exists for measuring and archiving high-temperature thermophysical property data for materials in the liquid, solid, and biphasic states. The important properties include emissivity, heat capacity, heat of fusion, melting temperature, density, surface tension, thermal diffusivity, and materials viscosity as a function of temperature and shear strain rate.

Comprehensive thermophysical properties, constitutive models for nonlinear behavior, intrinsic processing maps, and databases of microstructure-property relationships are needed for industrial process modeling as influenced by prior thermomechanical history. This database should represent a standardization of materials process design data that are certifiable in order for them to be useful to manufacturers. The database information currently used in process modeling is generally *typical data*, and the statistical assurance associated with such values is not known. Accurate error estimates in process model predictions require information regarding the statistical distribution of the input data.

Modeling of Abstraction of Downstream Constraints

As stated previously in this chapter, the design tradeoffs to be taken into account in concurrent engineering also must meet performance capability constraints such as inspectability, maintainability, and reliability. Inspectability during manufacturing and product service are not yet generally considered in the design process. However, recent difficulties with inspectability of aging aircraft (Achenbach and Thompson, 1991) have clearly indicated the need for incorporating inspectability at an early design stage.

Figure 4-2 shows the diagram of a concurrent engineering environment that links design to inspectability and the other downstream capability constraints, as well as to areas such as quality assurance, life-cycle costs, and materials and processes. At the present time, the only links that have been developed are the CAD/CAM links between design and manufacturing methods and processes.

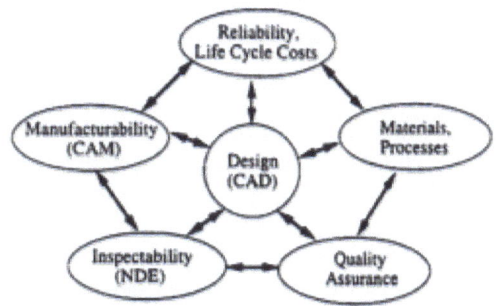

Figure 4-2 Concurrent engineering environment including inspectability. Reprinted courtesy of D. Thompson and L. Schmerr, Center for Nondestructive Evaluation, Iowa State University.

Computer models for other downstream capability constraints can provide key ingredients for implementing the complete concurrent engineering environment of Figure 4-2. In an early stage of the design process, models can be used to determine the role of such procedures as nondestructive evaluation (NDE) for in-process control of important parameters in the manufacturing process and for in-service use and in-the-field inspections. They also play an essential role in a damage-tolerant design philosophy and in questions of in-service reliability and life-cycle costs. Significant progress has been made in establishing NDE models and in building the other concurrent engineering links, such as through the joint National Institute of Standards and Technology, Iowa State University, and Northwestern University Program in Integrated Design, NDE, and the Manufacturing Sciences.

Measurement Modeling

The availability of a measurement model has many benefits. Numerical results based on a reliable model are very helpful in the design and optimization of efficient testing configurations. A good model is indispensable in the interpretation of experimental data and the recognition of characteristic signal features. The relative ease of parametric studies based on a measurement model facilitates an assessment of the probability of detection of anomalies. A measurement model is a virtual requirement for the development of an inverse technique based on quantitative data. If tested for accuracy by comparison with experimental data, it provides a practical way of generating a training set for a neural network or a knowledge base for a computer-aided system. Finally, and most importantly in the present context, these models can be incorporated into a concurrent engineering design process.

One of the most significant advances in nondestructive evaluation over the last decade has been the evolution of quantitative nondestructive evaluation (QNDE) from a conglomeration of empirical techniques to a well-defined field interdisciplinary science and engineering. In the course of this development, it has become well recognized that a fundamental approach to QNDE must be based on quantitative models of the measurement processes of the various inspection techniques. A model's principal purpose is to predict, from first principles, the measurement system's response to specific anomalies in a given material or structure (e.g., cracks, voids, distributed damage, corrosion, or deviations in material properties from specifications). Thus, a measurement model must include the configuration of probe and component being inspected and a description of the generation, propagation, and reception of the interrogating energy. For example, in the case of ultrasound as interrogating energy, this description requires computations of the transducer radiation pattern, refraction of the beam at the part's surface, the beam profile, and the propagation characteristics in the host material including effects of material anisotropy, attenuation, and diffraction losses. Detailed modeling of the field-flaw interactions that generate the measurement system's response function are also included, as well as information on material properties and other conditions that increase variability and add uncertainty to the measurement results. A well-constructed measurement model should be able to predict specific instrumental responses to any anomalies in complex materials and structures as well as to any *standard* flaws placed in various calibration blocks. The status of models for ultrasonics, eddy current methods, and radiographic techniques has recently been discussed by Gray et al. (1989).

Quantitative Nondestructive Evaluation

This section discusses QNDE as a measurement model and its application to a damage tolerant design philosophy and detection probability. The coupling of measurement models to CAD is also reviewed.

The load-bearing capacity of a structural system can conventionally be determined by applying increasingly larger loads until the structure fails. Such proof testing is part of the design process. Once a structure is in service, a proof test is obviously not a practical way to assess a part's condition. A feasible approach to obtaining strength information under in-service conditions is by using a QNDE technique, whereby a material or a structure is evaluated through interaction with some form of interrogating energy. Many forms of radiated energy have been used in QNDE (e.g., laser light, ultrasound, eddy currents, and x rays). Other techniques are based on the penetration of neutron and thermal waves. The QNDE approach includes the development of nondestructive measurement procedures to determine material properties and to detect flaws and other failure-related conditions. QNDE also encompasses the design of instrumentation, data processing, the use of measurement models, and the interpretation of data to determine whether a part should be rejected or a structural system should be repaired. QNDE procedures should be considered in the design stage as part of quality assurance, maintainability, and reliability analysis.

Fracture mechanics and failure mechanics have made great strides in the understanding and prediction of the integrity of structural components. For a component made of a material of known properties subjected to a given set of loads, it is possible to calculate the critical size of a crack at a specified location. A component is judged to be safe if the crack is smaller than a critical size and is not expected to grow to critical size during the service life or prior to the next inspection. Reliable methods must be available to detect and characterize cracks, including those of subcritical size. QNDE provides the technology to detect cracks (or more generally flaws) larger than the detectability limit and to determine location, size, shape, and orientation.

Damage-Tolerant Design

In a damage-tolerant approach, subcritical flaws just below the detection limit are assumed to exist at every fracture-critical location. As part of the analytical evaluation the following questions must be answered:

- What is the critical flaw size that will cause component failure when subjected to known service loads and temperature conditions?
- What are the driving forces causing crack growth?
- How fast will a subcritical crack grow under service load and temperature, and hence, how long can a component containing a subcritical flaw be safely operated in-service?
- What inspection must be performed to detect a crack before catastrophic failure of the component occurs?

Subcritical crack initiation and propagation occurs in high-stress areas and in locations where components contain material- and manufacturing-related inhomogeneities such as voids, inclusions, machining marks, or sharp scratches. Current programs require an inspection at half the time required for a potential crack to grow to critical size. The inspection is assumed to detect any flaw larger in size than a defined limit (Cowie, 1989).

A systematic approach to the overall inspection requirements of structures is required within a CAMSS system to advise the designer about potential problems with inspection and possible design alternatives. This approach should take into account the statistics of the occurrence of flaws, the crack growth mechanisms, and the various nondestructive detection techniques. The probability of detection of certain classes of defects also depends on the load, damage deterioration properties of the material, inspection intervals, human factors, and replacement and repair methodologies.

Probability of Detection

The implementation of a measurement model should be coupled to the concept of probability of detection (POD). This is a statistical representation of the probability that a given measurement system will be able to detect a specific flaw (or condition) in a given material or structure. It incorporates knowledge of the signal detected by the measurement system together with statistical information concerning flaw distributions, instrumental noise, and threshold levels. A POD curve shows the probability of a flaw's detection as a function of flaw size for a specific inspection technique. For an ideal technique, the POD of flaws smaller than a size predetermined by performance requirements and material properties is zero, while the POD for any flaw greater than this size is unity. In this case, there are neither false rejections of good parts nor false acceptances of defective ones. However, NDE techniques in practice are never as sharp and as discriminatory as indicated by the ideal curve. Thus, there are regions of uncertainty with false rejections and false acceptances.

Figure 4-3, which was taken from Gray and Thompson (1986), shows the results of simulating the ultrasonic POD of circular cracks at different depths below a cylindrical component surface and for two different scan plans. The plot on the left illustrates the use of the POD model to quantify the detection capability of an NDE system. For the specific parameters in that simulation, cracks that are otherwise identical have significantly different detectability levels depending on their depth below the surface of the part. This example illustrates the use of the POD model both for quantifying the capability of a flaw detection system and for suggesting improvements in either the system or its operation that can improve its capability.

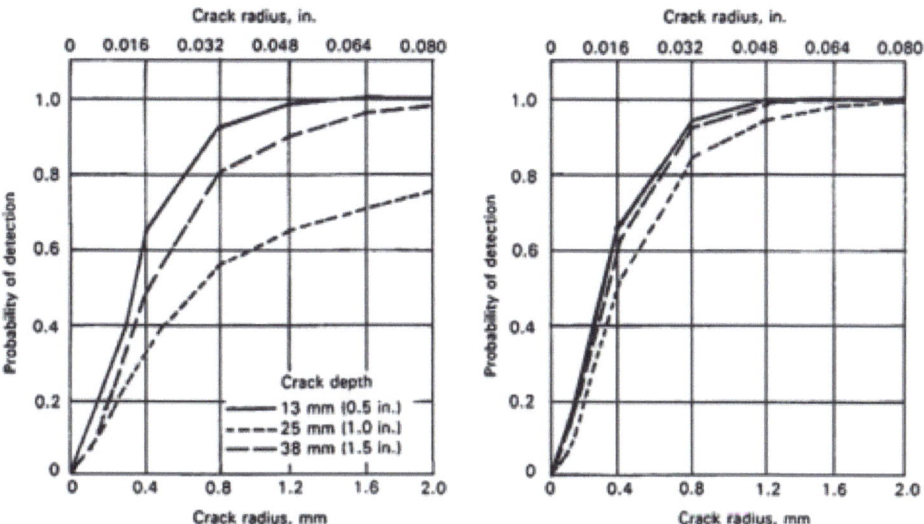

Figure 4-3 POD curves for two scanning plans. Source: Gray and Thompson, 1986.

Issues Concerning Implementing Modeling and Analysis Systems

If modeling and analysis systems in a CAMSS are to be useful and effective, future engineers must receive sufficient training in both the theory behind these systems and their application to the design process. A previous National Materials Advisory Board report entitled *Enabling Technologies for Unified Life-Cycle Engineering of Structural Components* stated that education in materials synthesis and processing is a barrier that must be mutually addressed by industry and U.S. institutions of higher learning (NRC, 1991). The time to introduce CAMSS and to perfect the skills in using the technique will be dependent on the availability of individuals with expertise in computer and materials science. Some engineers, such as manufacturing mechanical engineers, are highly skilled at using computers and doing FEA but do not have sufficient knowledge about the behavior of materials under processing conditions. In contrast, materials scientists and engineers have a better understanding of material behavior but lack sufficient training in the use of computers and computer-aided systems in manufacturing. Also, nondegreed technical personnel perform many crucial tasks throughout engineering and manufacturing that will be dependent on these new technologies. Institutions of higher learning will have to develop interdisciplinary programs led jointly by experts in materials science and engineering, design, and computer science if a CAMSS is to be properly implemented.

5

Conclusions and Recommendations

This report has discussed the structural engineering design process (Chapter 2), the vision for a CAMSS (Chapter 3), and the key materials-specific information technologies that could impact the development of a CAMSS (Chapter 4).

The committee has identified two main areas that are currently preventing the development of a CAMSS: (1) database and knowledge base design, implementation, instantiation, and management and (2) structural design modeling technologies. The barriers within these areas and the recommended R&D required to overcome these barriers are discussed in the first section of this chapter. General recommendations for government, industry, and university collaboration to forward the development of CAMSS are discussed in the second section of this chapter.

STRATEGIES FOR OVERCOMING BARRIERS

The committee identified two main areas that are currently preventing the development of a CAMSS: (1) database and knowledge base design, implementation, instantiation, and management and (2) structural design modeling technologies.

Database and Knowledge-Base Barriers

The design, implementation, instantiation, and maintenance of materials properties databases and knowledge bases are integral to the development of an effective CAMSS. For example, a design engineer cannot use a system if the underlying databases contain obsolete, extraneous, unverified, or incomplete information. The committee has found that the database and knowledge base area is currently inhibited by five barriers.

1. *Standardization of databases and knowledge bases*—Constructing databases and knowledge bases that contain the relevant information required for the design process and developing systems that locate and present this data are two difficult problems because of the amount of extraneous information available and the lack of standards in the content of databases and knowledge bases. *To overcome these barriers, the committee recommends that (1) standards and guidelines be developed for electronic data quality, capture, storage, analysis, and exchange (following the Computer-Aided Acquisition and Logistics Support and the Standard for the Exchange of Product approaches) and knowledge base content and construction; (2) CAMSS be designed to accept a variety of database taxonomies through the use of active, "intelligent" data dictionaries that aid the identification and conversion of the contents of different databases for use in the system; (3) links between materials databases and knowledge bases be improved and computer networks for materials-specific information communication be created (e.g., an electronic* Journal of Materials Selection in Structural Design, *a national materials bulletin board on Internet, or a linked network of worldwide materials data systems); and (4) electronic technical assistance be provided to design teams in electronic formats.*

2. *Status of knowledge capture*—Methods for knowledge capture are required to enhance the lessons-learned segment of CAMSS. These include establishing knowledge-representation taxonomies, technical context standards, and techniques to update and access this information rapidly. *To overcome this barrier, the committee recommends that (1) materials and computer scientists collaborate in the development of suitable knowledge-capture systems for use in*

CAMSS; (2) industry design teams be encouraged to establish electronic technical databases by electronic capture of all design discussions, decisions, and lessons learned in free text, spreadsheet, CAD standards, and other multimedia formats; and (3) industry design teams be encouraged to assign specific functions within the team to specialize, categorize, index, and filter the accumulated design knowledge base and locate and access other design knowledge bases.

3. *Diffuse responsibility for generating databases*—The issue of how to coordinate materials developers, component users, and materials societies to generate and integrate materials property databases requires resolution. Materials suppliers predominantly leave materials qualification programs to the user because of concerns that they will be held liable for system malfunctions caused by failures and that users will only employ materials that they themselves have qualified. Materials societies generally do not have the resources necessary for large projects. Component manufacturers typically only qualify materials for a given application and treat the data as proprietary. *To overcome this barrier, the committee recommends that (1) national team efforts of users, suppliers, materials societies, and standards organizations develop integrated material qualification programs that relate to design requirements and eventual use and (2) the resultant appropriate, independently verified data be made available in a national information infrastructure to provide a realistic, initial appraisal of the advantages of a material.*

4. *Disclosure of materials data*—In general, companies protect as proprietary all databases and knowledge bases that contain materials properties and production-related data, such as (1) state-of-the-art information, projections, or forecasts; (2) manufacturing labor standards, rates, and price data; and (3) weight, performance, and cost tradeoff data and criteria. *To overcome this barrier, the committee recommends that CAMSS be designed to assure that proprietary portions of databases and knowledge bases are fully protected.*

5. *Investment to maintain databases*—It is important that the information within a database be constantly monitored, verified, and updated to ensure that the best possible information is available. *To overcome this barrier, organizations must (1) assign the responsibility for maintenance of databases to a centralized function, such as a data administrator with domain experts identified to act as curators of the knowledge base, and (2) provide long-term support for database maintenance once the program is established.*

Structural Design Modeling Technology Barriers

Modeling in structural design will be an important component of any CAMSS both to provide valid details on which to base tradeoff decisions and to reduce reliance only on materials-properties databases. Modeling techniques are required for geometric reasoning, material responses on multiple scale levels, materials processing, manufacturing processing performance, product performance, and life-cycle issues such as inspectability. Modeling techniques will also be required that simulate new materials by successive extrapolation from the properties of existing materials or by calculation from first principles. The committee identified two barriers to the development of modeling.

1. *Optimization modeling*—As opposed to simply showing tradeoffs between design parameters input by users, modeling techniques will be required that can suggest modifications to optimize designs and manufacturing processes. Process optimization is an important ingredient of integrated product-process design and will be used more and more in the future as the industry fully adopts concurrent engineering to reduce manufacturing costs and converge on manufacturing solutions in a shorter time. To be useful, modeling must also be done rapidly and accurately, using normal design parameters and information from multiple knowledge bases. If modeling techniques are too slow, untrustworthy, or unable to access the proper information, they will languish. *To overcome these barriers, the committee recommends that (1) materials scientists and*

computer engineers from industry and university collaborate to develop advanced modeling techniques to reduce reliance on straight materials data, introduce expert knowledge, provide a credible basis for tradeoff decisions, and increase trust in CAMSS; and (2) materials scientists participate in basic and applied research that establishes links between materials models at several scales (e.g., atomic, molecular-crystal, cluster-grain size, polycrystal-aggregate, sub-structure, structure, and system).

2. *Cultural and educational barriers to implementing modeling and analysis technology*—The design process is traditionally a heuristic trial-and-error approach. Increased reliance on modeling techniques requires establishing confidence that the improved design solutions can be developed in a shorter time period. Current engineering programs do not stress the importance of training in either materials synthesis and processing or computer science. For modeling and analysis systems in a CAMSS to be useful and effective, future engineers must receive training in computer systems, modeling and analysis systems theory, and their application to the design process. *To overcome the cultural and educational barriers, the committee recommends that institutions of higher learning develop interdisciplinary programs led jointly by experts in materials science and engineering, design, and computer science that (1) expose student teams to basic approaches to computer-assisted concurrent engineering design systems in order to produce knowledgeable workers with a broad understanding of the science of processing, (2) train journeymen or master technologists to use this new technology to push acceptance of process modeling techniques to the shop floor, and (3) encourage younger faculty members to collaborate with colleagues in other departments (e.g., materials science, the traditional engineering fields, and computer science) on interdisciplinary design projects and computer-assisted concurrent engineering design systems.*

GENERAL CONCLUSIONS AND RECOMMENDATIONS

The areas inhibiting the development and implementation of CAMSS discussed above can only be overcome by a multipronged initiative with full participation and support by the IPDTs and materials and computer scientists and engineers in the government R&D agencies, universities, and industrial organizations.

The implementation of this vision will require (1) the development of significant demonstrations of CAMSS and disseminating the results; (2) the continued expansion of electronic storage of materials information; (3) the rapid adoption and application of developing methods of computer science and technology to enhance the representation of materials design knowledge; (4) the continued development of multilevel (atomistic to macroscopic) materials processing and constitutive behavior models that reliably predict performance and manufacturability at the scale of application; and (5) the implementation of methods to address inspectability, reliability, and maintainability.

Adherence to uniform computing and materials description standards in such programs is essential to the networked linking of individual tools into much larger design knowledge and support systems in the future. The committee recommends a higher level of communication among materials-specific information systems researchers and developers through a more formal electronic interchange of research information, network-linked use of computer-aided system tools, and access to electronic materials knowledge bases.

Recommendations specific to developers and users of CAMSS are

- *Government policy makers* should promote (1) the development of pre-competitive R&D programs that encourage industry, university, and government laboratories to leverage expertise and knowledge to reduce the time to develop, standardize, and implement product design support systems and materials-specific information technologies and (2) the use of the information super-highway as a means for expediting the sharing of technical information and memory among federal agencies, industries, and materials societies.
- *Government R&D organizations* (Department of Defense, Advanced Research Projects Agency,

National Aeronautics and Space Administration, National Institute of Standards and Technology) should promote database and knowledge base construction and standardization, design-knowledge tool demonstrations, and pilot projects as part of their future systems programs. These programs should integrate existing computer-aided system tools. *Two potential ways in which this might be accomplished are to provide (1) funding for demonstration programs with creative problem solving and design concepts to teams of university faculty and students composed of computer scientists, engineering design specialists, materials scientists, and cognitive psychologists and (2) financial incentives to industry for sharing materials property data where input to public and limited access materials knowledge bases can be controlled.*

- *Industries and universities* should be encouraged to collaborate in:
  1. developing and using well-defined standards for electronic information sharing to enable selective protection of organizational private data, company proprietary data, and industry restricted data from the public domain data;
  2. improving contact between researcher, designer, and supplier on design teams;
  3. increasing rate of generation, validation, and exchange of materials data;
  4. developing powerful programs for service life prediction of structural components from materials data, constitutive models, and in-service nondestructive testing;
  5. developing models of practical significance to product development;
  6. providing materials development data in machine readable electronic format;
  7. preparing standards for knowledge representation of materials information (e.g., properties tables, graphs, and pictorial descriptions of microstructures);
  8. publicizing success stories where experienced engineers select materials showing that proper representations together with reasoning examples will promote effective material computer-aided systems development; and
  9. developing an information base on available (network accessible) materials databases and computer-aided systems.

References

Achenbach, J.D., and D.O. Thompson. 1991. Towards quantitative nondestructive evaluation of aging aircraft. Pp. 1–14 in Structural Integrity of Aging Airplanes, S.N. Atluri, S.G. Sampath, and P. Tong, eds. New York: Springer-Verlag.

ACM (Association for Computing Machinery). 1993. Multimedia in the workplace. Communications of the Association for Computing Machinery 36(1).

ACM (Association for Computing Machinery). 1994. Special Issue on Intelligent Agents. Communications of the Association for Computing Machinery 37(7).

Allen, B.P. 1994. Case-based reasoning: Business applications. Communications of the Association for Computing Machinery, 37(3):41–42.

Ambler, E. 1985. Engineering Property Data: A National Priority. ASTM Standardization News, August:46–50.

Annis, C.G., M.C. Van Wanderham, J.A. Harris, and D.L. Sims. 1981. Gas turbine engine disk retirement-for-cause: An application of fracture mechanics and NDE. Journal of Engineering for Power 103(1):198–200.

Ashby, M.F., and D.R.H. Jones. 1980. Engineering Materials: An Introduction to Their Properties and Applications. New York: Pergamon Press.

Boeing Commercial Airplane Group. 1991. Presentations to the Committee on Application of Expert Systems to Materials Selection During Structural Design, Seattle, December 10–11.

Brathwaite, K.S. 1988. Analysis, Design, and Implementation of Data Dictionaries. New York: McGraw-Hill.

Burte, H.M., and C.L. Harmsworth. 1989. Database R&D for unified life-cycle engineering. Pp. 197–199 in Computerization and Networking of Materials Data Bases, J.S. Glazman and J.R. Rumble, eds. ASTM Special Technical Publication 1017. Philadelphia, Pennsylvania: ASTM.

Clocksin, W.F., and V.S. Melish. 1981. Programming in PROLOG. New York: Springer-Verlag.

Cowie, W.D. 1989. Fracture Control Philosophy. Pp. 666–673 in Metals Handbook: Nondestructive Evaluation and Quality Control, Vol. 17. Metals Park, Ohio: ASM, International.

Coyne, R.F. 1991. ABLOOS: An Evolving Hierarchical Design Framework. Ph.D. dissertation. Department of Architecture. Carnegie Mellon University, Pittsburgh, Pennsylvania.

Coyne, R.F., and U. Fleming. 1990. Planning in design synthesis: Abstraction-based LOOS. Pp. 91–111 in Proceedings of the Fifth International Conference on Applications of Artificial Intelligence in Engineering, Vol. 1, J.S. Gero, ed. Southampton, England: Computer Mechanics Publications.

Davis, R., H. Shrobe, and P. Szolovits. 1993. What is knowledge representation? AI Magazine 14(1):17–33.

DOD (Department of Defense). 1986. Military Standardization Handbook: Metallic Materials and Elements for Aerospace Vehicle Structures, MIL-HDBK-5E. Washington, D.C.: Department of Defense.

Eberhard, E., and D.D. Vvedensky. 1987. Localized grain-boundary electronic states and intergrannular fracture. Physics Review Letters 58 (1).

Famili, A., D.S. Nau, and S.H. Kim, eds. 1992. Artificial Intelligence Applications in Manufacturing. Cambridge, Massachusetts: Massachusetts Institute of Technology Press.

Gray, T.A., and R.B. Thompson. 1986. Use of models to predict ultrasonic NDE reliability. Pp. 911–918 in Review of Progress in Quantitative Nondestructive Evaluation, Vol. 5, D.O. Thompson and D.E. Chimenti, eds. New York: Plenum Press.

Gray, J.N., T.A. Gray, N. Nakagawa, and R.B. Thompson. 1989. Models for Predicting NDE Reliability. Pp. 702–715 in Metals Handbook: Nondestructive Evaluation and Quality Control, Vol. 17. Metals Park, Ohio: ASM International.

Gruber, T., C. Baudin, J. Boose, and J. Weber. 1991. Design rationale capture as knowledge acquisition: Tradeoffs in the design of interactive tools . Pp. 3–12 in Machine Learning: Proceedings of the Eighth

REFERENCES

_____ International Workshop, L.A. Birnbaum and G.C. Collins, eds. San Mateo, California: Morgan Kaufmann.

Gursoz, E.L., and F.B. Prinz. 1990. A point set approach in geometric modeling. Pp. 73–87 in Proceedings of the International Symposium on Advanced Geometric Modeling for Engineering Applications, R.L. Krause, and H. Jansen, eds. New York: Elsevier Science Publications.

Hadcock, R.N. 1985. X-29 Composite Wing. Presented at the American Institute of Aeronautics and Astronautics Symposium on the Evaluation of Aircraft/Aerospace Structures and Materials, Air Force Museum, Dayton, Ohio, April 24–25.

Hadcock, R.N. 1986. Structures and Materials Technology. Presented at the Grumman Aircraft Systems Division Forum, Bethpage, New York, August 26.

Harmon, P., and D. King. 1985. Expert Systems. New York: John Wiley & Sons.

Harrison, R.J., and I. Hulthage. 1987. Specification of knowledge base systems. In Proceedings of the 34th Sagamore Army Materials Research Conference. Watertown, Massachusetts: Materials Technology Laboratory.

Hayes-Roth, F., and N. Jacobstein. 1994. The state of knowledge based systems. Communications of the Association for Computing Machinery 37(3):27–39.

IEEE (Institute of Electrical and Electronic Engineers). 1993. IEEE Spectrum Special Report on Interactive Multimedia 30(3).

Kaufman. J.G. 1986a. Standardization for materials property databases and networking. ASTM Standardization News. February:28–33.

Kaufman, J.G. 1986b. Technical challenges and the status: the national materials property data network. Pp. 159–166 in The Material Property Data: Applications and Access, Vol. 1. New York: American Society of Mechanical Engineers.

Kaufman, J.G. 1988. An Intelligent Knowledge System for Selection of Materials for Critical Aerospace Applications (IKSMAT). Final Technical Report, U.S. Air Force Contract F33615-87-C-5305. Columbus, Ohio: Department of Defense.

Kim, S.H. 1990. The Essence of Creativity: A Guide to Tackling Difficult Problems. New York: Oxford University Press.

Kiridena, V., A.B. Chaudhary, and J. Gunasekera. 1989. Analysis of large deformation 3-D metal forming analysis with intermediate remeshing. In Numerical Methods in Industrial Forming Process, E.G. Thompson, ed. Boston, Massachusetts: A.A. Balkema.

Klahr, D., P. Langley, and R. Neches, eds. 1987. Production System Models of Learning and Development. Cambridge, Massachusetts: Massachusetts Institute of Technology Press.

McCarthy, J. 1968. Programs with common sense. In Semantic Information Processing, M. Minsky, ed. Cambridge, Massachusetts: Massachusetts Institute of Technology Press.

Minsky, M., ed. 1968. Semantic Information Processing. Cambridge, Massachusetts: Massachusetts Institute of Technology Press.

Mobley, C.E. 1992. Quality Castings: CT 17 Castability Maps. Kettering, Ohio: Edison Materials Technology Center.

NRC (National Research Council). 1983. Materials Properties Data Management—Approaches to a Critical National Need. National Materials Advisory Board, NRC. Washington, D.C.: National Academy Press.

NRC (National Research Council). 1989. Materials Science and Engineering for the 1990s: Maintaining Competitiveness in the Age of Materials. National Materials Advisory Board, NRC. Washington, D.C.: National Academy Press.

NRC (National Research Council). 1991. Enabling Technologies for Unified Life-Cycle Engineering of Structural Components. National Materials Advisory Board, NRC. Washington, D.C.: National Academy Press.

NRC (National Research Council). 1993. Commercialization of New Materials for a Global Economy. National Materials Advisory Board, NRC. Washington, D.C.: National Academy Press.

Parsaye, K., M. Chignell, S. Khoshafian, and H. Wong. 1989. Intelligent Databases: Object-Oriented, Deductive, Hypermedia Technologies. New York: John Wiley & Sons.

Piatetsky-Shapiro, G., and W.J. Frawley, eds. 1993. Knowledge Discovery in Databases. Cambridge, Massachusetts: Massachusetts Institute of Technology Press.

Rappaport, A., and R. Smith. 1991. Innovative Applications of Artificial Intelligence, Vol. 2. Cambridge,

REFERENCES

Massachusetts: Massachusetts Institute of Technology Press.

Reynard, R. 1987. Computerized materials data: The SDI view. In Proceedings of Workshop on Computerized Aerospace Materials Data, J.H. Westbrook and L.R. McCreight, eds. New York: American Institute of Aerospace and Aeronautics.

Richmond, O. 1992. Constitutive Model-Based Material Product Design. Presented to the National Materials Advisory Board Committee on the Application of Expert Systems to Materials Selection During Structural Design, April 27, Washington, D.C.

Robinson, S. 1987. The Spang Robinson Report, Vol. 3, Number 3. Palo Alto, California: Spang Robinson.

Robinson, J.A., and E.E. Sibert. 1981. The LOGSLISP User's Manual. Technical Report. Syracuse, New York: School of Computer and Information Science, Syracuse University.

Rumelhart, D.E., B. Widrow, and M.A. Lehr. 1994. The basic ideas in neural networks. Communications of the Association for Computing Machinery 37(3):87–92.

Schurr, H., and A. Rappaport. 1989. Innovative Applications of Artificial Intelligence, Vol. 1. Cambridge, Massachusetts: Massachusetts Institute of Technology Press.

Smith, R., and A.C. Scott. 1991. Innovative Applications of Artificial Intelligence, Vol. 3. Cambridge, Massachusetts: Massachusetts Institute of Technology Press.

Thompson, D.O. 1993. Tools for a NDE Engineering Basis: Flight-Vehicle Materials, Structures and Dynamics Technology—Assessment and Future Directions. New York: American Society of Mechanical Engineers.

Weber, J.C., B.K. Livezey, J.G. McGuire, and R.N. Pelavin. 1991. Integrating specialized representations for spreadsheet-like design. Pp 504–511 in Proceedings of Fourth International Conference on Industrial and Engineering Applications of Artificial Intelligence. Tulahoma, Tennessee: University of Tennessee Space Institute.

Weiss, V., and D.W. Aha. 1984. Materials selection with logic programming. Pp. 506–518 in Proceedings of the Society for the Advancement of Material and Process Engineering Symposium on Technology Vectors, Vol. 29. Covina, California: Society for the Advancement of Material and Process Engineering.

Wertz, C.J. 1989. The Data Dictionary Concepts and Uses, Second Edition. Wellesley, Massachusetts: QED Information Sciences.

Whitney, D.E., J.L. Nevins, T.L. De Fazio, R.E. Gustavson, R.W. Metzinger, J.M. Rourke, and D.S. Seltzer. 1988. The strategic approach to product design. Pp. 200–223 in Design and Analysis of Integrated Manufacturing Systems, W.D. Compton, ed. Washington, D.C.: National Academy Press.

Widrow, B., D.E. Rumelhart, and M.A. Lehr. 1994. Neural networks: Applications in industry, business, and science. Communications of the Association for Computing Machinery 37(3):93–105.

Winner, R.I., J.P. Pennell, H.E. Bertrand, and M.M. Slusarczuk. 1988. Role of Concurrent Engineering in Weapons Systems Acquisition. IDA Report R-338 (unclassified). Alexandria, Virginia: Institute for Defense Analysis.

REFERENCES

Appendix A:
Glossary of Acronyms

AI	Artificial intelligence
CAD	Computer-aided design
CAD/CAM	Computer-aided design and manufacturing
CAM	Computer-aided manufacturing
CAMSS	Computer-aided materials selection system
DBT	Design build team
DMM	Dynamic materials modeling
FEA	Finite element analysis
FEM	Finite element method
IKSMAT	Intelligent knowledge system for selection of materials for critical aerospace applications
IPDT	Integrated product development team
KIDS	Knowledge based integrated design system
NDE	Nondestructive evaluation
OEM	Original equipment manufacturer
POD	Probability of detection
QNDE	Quantitative nondestructive evaluation
R&D	Research and development
SME	Small-to-medium enterprise

APPENDIX A:

Appendix B:
Case Studies Reviewed by the Committee

A Materials Selection Expert System for Corrosive Aqueous Environments, V. Weiss, Syracuse University, New York

Design for High-Speed Civil Transport Applications, P. Rimbos, HSCT Structures Technology Development, The Boeing Company, Seattle, Washington

The Role of Materials Engineers in Hardware Design, T. Richardson, 777 Airplane Development, The Boeing Company, Seattle, Washington

Allowables Perspective of Materials Selection in the Design Process, B.F. Backman, The Boeing Company, Seattle, Washington

Design Build Team (DBT) Approach to Product Development, H. Shomber, Design 777 Division, The Boeing Company, Seattle, Washington

DBT Experiences in Current Hardware Programs, A. Falco and T. Lackey, Design Engineers 777 Empennage, The Boeing Company, Seattle, Washington

Computer-Aided Design Tools Currently in Practice, T.S. Kaczmarek, Artificial Intelligence, General Motors Corporation, Warren, Michigan

Quality Assurance Perspective of DBT New Airplane Program, B. Das, 777 Quality Assurance/Quality Engineering, The Boeing Company, Seattle, Washington

NASA Funded ACT Program: COINS and COSTADE Design Tools, L. Ilcewicz, Advanced Technology Composite Aircraft Structures, The Boeing Company, Seattle, Washington

Composite Materials Selection: A Suppliers View Point, J. Hendrix, Hercules Incorporated Washington, DC

Constitutive Model-Based Material Product Design, O. Richmond, ALCOA, Alcoa Center, Pennsylvania

Intelligent Processing and Materials Modeling, W. Barker, ARPA, Department of Defense, Arlington, Virginia

Design Knowledge Capture for Corporate Memory, J. Boose, Boeing Advanced Technology Center, Computer Science, The Boeing Company, Seattle, Washington

Design Cost Model Studies for Advanced Composite Fuselage, G. Swanson, Advanced Technology Composite Aircraft Structures, The Boeing Company, Seattle, Washington

Material Selection for Commercial Airplanes, A. Miller, Beoing Materials Technology, The Boeing Company, Seattle, Washington

Integration of Material Service/Life-Cycle Considerations in the Design Process, A. Miller, Boeing Materials Technology, The Boeing Company, Seattle, Washington

Artificial Intelligence in Design of Materials and Structures, F. Crossman, Lockheed Palo Alto Research Laboratory, Palo Alto, California

Life Prediction Data/Methodology for a MSES, J. Schreurs, Westinghouse Electric Company, Westinghouse Science and Technology Center, Pittsburgh, Pennsylvania

Knowledge-Engineering and Representation Methods for Material Selection, T.S. Kaczmarek, Artificial Intelligence, General Motors Corporation, Warren, Michigan

Superconductor Search: An Expert System for the Development of High-Temperature Superconductors, J. Schreurs, Westinghouse Electric Company, Westinghouse Science and Technology Center, Pittsburgh, Pennsylvania

Design Specifications for Knowledge Based Systems for Materials Design, I. Hulthuse, Robotics Institute, Carnegie-Mellon University

An Intelligent Knowledge System for Critical Aerospace Systems, W.M. Griffith, Wright Patterson Air Force Base, Ohio

Advanced Materials Database System, W.M. Griffith, Wright Patterson Air Force Base, Ohio

Expert System for Failure Analysis of Aircraft Metallic Materials and Ground Support Equipment, W.M. Griffith, Wright Patterson Air Force Base, Ohio

A Database Management System for MMCs, W.M. Griffith, Wright Patterson Air Force Base, Ohio

APPENDIX B:

Advanced Ceramics Information Systems, R.G. Munro, Ceramics Division, National Institute of Standards and Technology, Gaithersburg, Maryland

M/Vision Materials Selection System, D. Marinaro, Software Engineering Group, PDA Engineering, Costa Mesa, California

Design Information System, F. Crossman, Lockheed Palo Alto Research Laboratory, Palo Alto California

Knowledge Based Integrated Design System (KIDS), H.L. Gegel, Director, Processing Science Division, Universal Energy Systems, Dayton, Ohio

An Intelligent Knowledge System for Selection of Materials for Critical Aerospace Applications (IKSMAT), J.G. Kaufman, Vice President of Technology, Aluminum Association, Washington, D.C.

Appendix C:

Review of Selected Knowledge-Representation Techniques and Tools

Expert system implementations employ many different knowledge-representation techniques and tools. Each technique provides an abstraction that is useful in describing some aspect of expert behavior or an improved implementation of an abstraction concept. Just as words, numbers, graphs, and sketches are different but useful abstractions, the techniques described in this appendix are various ways of describing relationships and reasoning. Tools are implementations of knowledge-representation techniques. This appendix reviews several of the currently used representation technologies and tools discussed in the report and is not meant to be exhaustive. References are provided so that interested readers can further explore the techniques discussed here, as well as many others.

CASE-BASED REASONING

Case-based reasoning is a method for decision making based on the retrieval and adaptation of prior recorded cases. Tool functionality can range from retrieval, which only finds relevant cases in response to a user's input, to analogical reasoning, which finds and adapts a prior solution to the current situation. As such, case-based systems provide at least a primitive form of learning. Commercial tools for associative retrieval and case-based retrieval are available and significant applications are beginning to emerge.

CONSTRAINT-BASED REASONING

In constraint-based reasoning, knowledge is encoded as constraints that express qualitative or quantitative relationships between design parameters. Various algorithms exist to provide varying support of constraints ranging from violation detection, to enforcement, to propagation, to satisfaction.

When reasoning about constraints, the expert system must decide which constraints are relevant to the problem and then interpret them. Constraints and constraint reasoning can support design analysis by identifying problem areas. During design synthesis, constraints can be exploited to propose a solution that is acceptable within the problem domain, provided that the problem is not overly constrained. As an aid to search, constraints can be used to confine the search space. Systems that symbolically solve mathematics problems have been investigated since the early days of artificial intelligence (McCarthy, 1968; Minsky, 1968). Constraint reasoning is a more recent technology that has evolved in several different styles. In particular, reasoning about geometric invariance is critical to spatial reasoning. Linking geometric reasoning to symbolic reasoning will be critical for expert system technology in material selection.

Constraint-based algorithms vary in complexity as support ranges from detection of violations through satisfaction. Violation detection can be and has been done with a variety of rule-based languages as well as procedural code. Constraint propagation is available in several commercial products. Constraint satisfaction or solution is an active research area, although some algorithms with limited capabilities are available and in use.

ACTIVE DATA DICTIONARIES

Data dictionaries are organized references to data contained in other programs, systems, databases, or collections of files. Whereas databases store and process ordinary data about objects, data dictionaries contain data about data, or metadata. Active data dictionaries are used to coordinate and support data retrieval and analysis between different systems or databases. Although the implementation of active data dictionaries is predominantly a research area, some limited capabilities are currently available and in use as part of database systems.

DATABASES

Simply defined, databases store information according to a specified schema. Relational databases are commonly used today, and they store data like that represented by tables in reference works. Database management systems are an important component of most expert systems. They support dynamic factual recall, updating, and user access control, which may be thought of as a form of intelligent behavior. Furthermore, many of the recent advances in state-of-the-art database management systems have incorporated advanced concepts from expert database systems.

Relational electronic databases are organized as rigid tables, where each record of the database is assigned an equal number of fields, each containing a specified type of entry. Such rigid formatting is no longer necessary with the flexibility afforded by implementations such as association lists or structures and even arrays of structures. Free-format databases can save storage space when most records in the database contain entries only for a few of the many possible fields. Free-format databases are also preferred when the database is unstable and updated frequently, not only by adding records but also by adding fields or by modifying the requirements on a field. However, rigid formatting saves space and query time when most records of the database contain values for the same number of fields and allows nonprocedural queries to be made that automatically link together multiple tables.

Modern database packages are so versatile and easy to use that the materials scientist hardly needs to worry about database formats as long as the data are well-defined quantities or arbitrary text. However, analysis must be done in the design of a database or knowledge base. Factors to be considered include expected query scenarios and unusual data sets (e.g., default values, multiple entries, data ranges, incidental information such as warnings and comments, derived values, constraint ranges, quality, unit conversions, and educated guesses for missing entries).

Most expert system applications involve extracting information from existing databases. Some existing database packages may be able to handle certain types of reasoning about data within the database structure itself, but in most cases the knowledge engineer must either select a different form of knowledge representation to support reasoning requirements or design a new database to handle special problems. New technology for knowledge discovery in databases shows promise for making the information more useful in the construction of more advanced reasoning systems (Piatetsky-Shapiro and Frawley, 1993; see "Active Data Dictionaries"). Object-oriented databases are adding the security and access facilities of databases with the flexibility of artificial intelligence data structures and may overcome performance limitations to be widely used in the future.

A strength of computers is the ability to retrieve items stored in a database for use in other forms of knowledge representation or for display to inform the user. However, the quantity of data available for most advanced materials is inadequate for the types and depth of analyses needed for knowledge base systems, including statistically based design values. Database technology has outstripped the effort to build and distribute reliable data.

FUZZY LOGIC

Fuzzy logic is a method for dealing with the inherent ambiguities in concepts and an attempt to build a formal logic for plausibility. Fuzzy logic does not deal with probabilities but, rather, with the type of reasoning people use when faced with inconclusive or contradictory evidence. Fuzzy logic involves four basic elements: (1) schemes to convert stimuli signals to strength of belief, (2) simple rules expressed in logical terms, (3) algorithms for computing strength of belief for the conclusion of rules, and (4) output functions to convert the belief in conclusions to a control signal. Unlike the rules found in typical expert systems, which are complex and designed to support deep logical chains, fuzzy logic rules are very simple and do not involve the combination of conclusions to infer other conclusions.

Recent computing advances have led to successful applications of fuzzy logic in areas ranging from manufacturing controls to consumer electronic products. The main advantage of fuzzy logic is that it provides a simple formulation of simple reasoning processes. The simplicity of the reasoning restricts the application of fuzzy logic to narrowly scoped problems, however.

APPENDIX C:

GEOMETRIC AND MICROSTRUCTURAL INFORMATION REPRESENTATION

There are three levels of geometric modeling common in industry. One-dimensional (wireframe modeling), two-dimensional (surface modeling), or three-dimensional (solid modeling) analytical elements are used in constructing spatial representations. Figure C-1 displays the essential difference in the representational domain of each of the three levels of modeling. It should be clear that solid modeling has the most explicit representation of the objects in real-world and includes the lower levels of representation. This implicit embodiment, however, does not necessarily mean that any solid modeling approach can uniformly manipulate entities of the lower dimensions as well as solids.

Several taxonomies for solid modeling have been proposed, but one way of categorizing most of the existing solid model approaches is to perceive them in three classes:

- cell-based representations;
- constructive solid geometry representations; and
- surface boundary representations.[1]

Two distinct approaches in the category of cell based representations are the cell enumeration technique and the octree approach. In both of these cases, a solid is defined as a union of a selection of space-based cubical volumes. In constructive solid geometry schemes, objects are obtained by combining a set of solid primitives with boolean operators. In surface boundary representations, the objects are represented by their enclosing shell.

HYPERDOCUMENTS

Hyperdocuments are multimedia files in which the pieces of the documents are linked to one another to capture important relationships between concepts presented in the documents. Several commercial systems have led to numerous successful applications. The capturing of the relationships between objects acts as a primitive form of semantic network (see "Objects and Taxonomies").

MACHINE LEARNING

Machine-learning techniques allow a system to acquire knowledge automatically. Some simple techniques have been successfully applied and are commercially available, such as in the areas of case-based reasoning and neural networks, but most are still in the research phase.

MATHEMATICAL RELATIONS

Mathematical relations are equations, inequalities, approximations, and iterations that designers use to determine the properties of materials under certain conditions. Mathematical relations appear in exactly the same format in electronic knowledge bases as they do in books. The great advantage that computers have, however, is that they can actually compute values using equations, whereas books can only describe how to compute the values. One disadvantage of computers is that the covert use of equations to represent knowledge can be dangerous. Many equations, particularly in the materials and design fields, are based on approximations that are only valid for one application or within a specific range of variables. Good knowledge representation demands the existence of a mechanism, commonly referred to as "explanation," to permit the user to inspect the equations that are invoked and the assumptions that are inherent in the choice of the equations used in the calculation to avoid potential problems.

When dealing with mathematical relations, the knowledge engineer must decide whether these relations are to be used as symbolic expressions or simple computations. If they are to be used as computations only, then conventional programming languages can be used to perform the

---

[1] Surface boundary representation of a solid refers to the closing surface of an otherwise open solid. Surface modeling, in general, may contain surfaces that do not necessarily enclose solids. An example is two surfaces that may interact but no not form an enclosed shell.

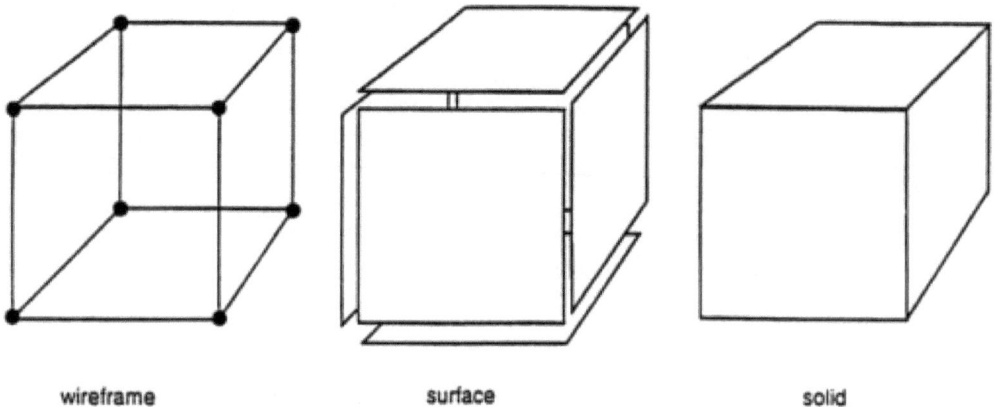

Figure C-1 The differences between wireframe, surface model, and solid model representational domains (Source: Gursoz and Prinz, 1990; Reprinted courtesy of Elsevier Science Publications).

arithmetic. Frequently, engineering problems require using mathematical relations as symbolic expressions of constraints, however. All constraints are not typically thought of as mathematical relationships and an expert system reasoning about constraints must also admit constraints expressed in a more logic-based formalism (see "Constraint-Based Reasoning").

NEURAL NETWORKS

Neural networks attempt to mimic brain-like systems via simplified mathematical models. Researchers have found that simple mathematical stimulus-response equations can be used to simulate the behavior of neurons in the brain. Like the brain, the most basic processing unit of neural networks is the neuron, which is characterized by "an activity level (representing the state of polarization of a neuron), an output value (representing the firing rate of the neuron), a set of input connections (representing synapses on the cell and its dendrite), a bias value (representing an internal resting level of the neuron), and a set of output connections (representing a neuron's axonal projections)" (Rumelhart et al., 1994). Neural networks analyze data by mapping input data into output patterns based on maps produced by previous runs.

A major advantage of neural networks is that the simple mathematical representation lends itself to learning algorithms. Using feedback, these algorithms adjust the set coefficients used to reinforce and combine stimuli to minimize an error score. Neural network learning algorithms require very large training sets and typically work best when the network connectivity has been properly organized in advance by an expert. Successful applications have been developed mainly in pattern recognition.

There are two main drawbacks with neural networks, however. In addition to requiring large training sets, neural networks do not have strong mechanisms for explaining the results of a computation. The latter problem is particularly troublesome in areas of engineering where tractability of design decisions is a requirement, such as in the design of products that affect public safety.

While many of the concepts of neural networks have been investigated for quite some time, this technology is in its early stages of application. Applications have only now become feasible because of low-cost computing developments.

OBJECTS AND TAXONOMIES

Objects and taxonomies are knowledge base tools that allow programmers to represent knowledge of physical or conceptual entities with many attributes in an abstract manner that mimics the way people organize knowledge about concepts and classes of objects. Objects and taxonomies are probably the most general and flexible form of knowledge-capture scheme available. They can handle databases, mathematical relations, rules, and anything that can be classified, including design features such as shapes and colors. Object-oriented programming requires a different software design approach than conventional programming and is still evolving.

There are many concepts that have been explored in artificial intelligence and programming language research that are similar to objects and taxonomies, some of which are commercially available in many forms as well as embedded in knowledge base engineering tools. Abstract data types, frames, schema, relational tables, and semantic networks are the most commonly referred to variants of the technology. All of these provide a means to describe facts and meaningful relationships between facts. They differ from data types found in conventional programming languages and databases in the expressive power regarding relationships. The price paid for this is less efficient programming and difficulty in providing shared access to data. Objects and abstract data types provide an additional benefit to the programmer or knowledge engineer by associating the processing or functional elements of the implementation with the kinds of data that the functions can manipulate. Thus, they provide more structure and understandability.

One disadvantage of maximizing generality and flexibility in a system is that a great deal of expertise is usually required to operate the system, so that the user effectively becomes a computer specialist as well as a materials scientist or design specialist.

REASONING WITH UNCERTAINTY

Reasoning with uncertainty is a knowledge-representation technique for combining contradictory, incomplete, or inconclusive knowledge. This is not the same as fuzzy knowledge or fuzzy logic, however. Expert systems can accommodate uncertainty by several approaches, including maintaining multiple problem formulations, qualitative methods, and quantitative methods involving uncertainty measures. Many applications have used some form of uncertainty logic.

RULE-BASED REASONING

Rules are representations of knowledge about which patterns of information experts use to make decisions and what are the decisions that follow. Rule-based reasoning provides automatic combination of rules to chain to a conclusion. One popular way to represent knowledge is the "if-then" rule. A rule can formally be represented as the logical relation:

$$p \rightarrow q$$

p represents a set of conditions or premises, and q represents a set of consequences or conclusions. Many different algorithms have been developed to implement and support the basic notion of rule-based reasoning. The differences between various approaches are in the domain of knowledge-engineering. For example, forward chaining rules facilitate programming synthesis, while backward chaining rules are more suited for analysis or search.

Rules are well suited for the type of reasoning that can typically be represented by a tree or a flow diagram. Rules typically represent reasoning about facts and data rather than the facts or data themselves (i.e., metadata). Expert systems based on rules include an implementation of an algorithm that governs what the rules can do, when they are activated or triggered, and what order of priority they are checked and executed. The software component controlling the rules is commonly referred to as an inference engine, since it controls the inferences of the system. Knowledge about materials can usually be stated in the "if-then" form. Rule-based knowledge representations can also handle limited forms of uncertain reasoning, such as by adding or subtracting confidence while appraising a hypothesis or by providing mechanisms to handle alternative lines of reasoning.

Many commercial tools are available that provide forward or backward chaining or both types of rules.

Rule-based tools are often characterized as expert system shells. Many successful applications have been developed in combination with other tools (e.g., objects). The main advantages of software packages that represent knowledge in rule form are that they allow the user to inspect the rules in near-natural language and provide an explanation of why a decision was made. Although it is easy for a human expert to understand a rule about material properties and to judge whether the rule is acceptable (a definite plus when one needs to know what knowledge has been brought to bear), knowledge engineers tend to clutter their rules with computing tricks that ultimately make reading, managing, modifying, or updating the rules by the user extremely difficult. This practice has led to unfair criticism of the underlying technology. Thus, rules should be used as appropriate in conjunction with other knowledge-representation forms.

SPATIAL SYNTHESIS AND LAYOUT

Conceptual layout is usually one of the first steps in creating a structure. These structures may be in the electrical, mechanical, architectural, or microstructural domain. The nature of the problem of formulating a layout is discussed, and some of the emerging computer technologies are described that can either automatically or with user guidance synthesize structures in two or three dimensions. The earlier described geometric representations can be used to implement such algorithms.

Layout design deals with many of the complex issues that typically arise in the design of artifacts that have to satisfy specified constraints and are composed of parts that have shape and occupy space. A large (potentially infinite) number of location and orientation combinations are available for placing any single object. In each combination, design objects interact in intricate ways through their shapes, sizes, and the spatial or topological relations that exist between them. These characteristics also interact in complex patterns with multiple performance criteria or functional attributes demanded of the artifact being designed. Layout design decisions must simultaneously satisfy global requirements (e.g., usage of space) and local requirements (e.g., adjacencies between pairs of objects with certain microstructures, as required in the design of sliding components); acceptable spatial arrangement often exhibits a complex pattern of tradeoffs.

For these reasons, there is no known direct method that is guaranteed to produce feasible solutions without iterations of trial and error for most application domains. Some amount of exploration of the structure, formulation of the layout task, and searching for candidate solutions is required. However, due to cognitive limitations, human designers do not have the capability for making systematic explorations of alternative arrangements. This shortcoming in human performance has motivated numerous attempts to apply computational methods to layout. What is desired is a structured method for producing multiple alternatives, each of which embodies tradeoffs that can be understood, justified, and indicative of a range of possible variations within which optimization can take place.

Attempts to arrive at such a method confront the challenges mentioned above. Consequently there is a long history of attempts to develop an effective, "closed-solution" computational-based method reflecting a variety of representations, system architectures, and planning strategies for layout design. Finding an effective representation to support the efficient generation and evaluation of design alternatives has been a difficult undertaking and has dominated the evolution of the field. The representation must support the creation of a space of possible designs by capturing meaningful differences between design alternatives at a manageable level of detail (or abstraction). Layouts for a given design problem are typically very large; therefore, the representation must allow for the employment of effective planning and search strategies to enable reasonable examination of the best alternatives (e.g., through the evaluation of partial solutions and the incremental specification of designs).

The layout operating system (LOOS; Coyne and Flemming, 1990), for example, enables the systematic generation of layout alternatives and their evaluation over multiple performance criteria. The system utilizes a graph-based representation that separates topological issues (spatial relations between objects) from metrical issues (dimensions and dimensional positions of objects) in layout. The representation uses basic spatial relations (i.e., above, below, to the right of, and to the left of) to define the structure or topology of a layout as a set of relations between pairs of rectangles. It represents this structure formally through an arc-colored directed graph, the vertices of which represent the rectangles in a layout and the arcs of which represent the spatial relations between the rectangles. Figure C-2 shows an example in which solid arrows indicate above/below relations, dashed arrows indicate left/right relations, and E represents the minimum enclosing rectangle that is above, to the right of, to the left of, and below all other rectangles in the layout. Using this representation, a set of rules or operations are defined that can generate all possible arrangements of rectangles in a plane by insertion of one rectangle at a time. The layouts produced by the LOOS are loosely packed arrangements of rectangles (e.g., the rectangles are nonoverlapping and need not fill the surrounding rectangle). Therefore, the approach is general enough to encompass a broad class of layouts and is useful over a wide range of domains. These rectangular arrangements are given meaning as layouts in a particular domain by attributing the layout objects or components from the domain to respective rectangles. In addition, tests or performance requirements for the layout are attached to these objects enabling the layouts produced to be comparatively evaluated. Those that fail requirements may be discarded, while those that show promise can be further developed.

Laying out abstract objects does not make a design; it only gives a spatially feasible configuration of the objects considered. The next step involves incorporating all detailed features of the design, both geometric as well as nongeometric ones. To facilitate this step, it is convenient to introduce a formal language with a grammar to express the intentions of where to generate what entity in what shape and size, and to determine what other nongeometric entity should be assigned to it.

Solids can be described through the surface boundary representation as previously introduced. Boundary solid grammar provides a means of generating complex models of rigid solid objects. Solids are represented by their boundary elements (i.e., vertices, edges, and faces with coordinate geometry associated with the vertices). Labels may be associated with any of these elements. Rules match conditions of a solid or collection of solids and may modify them or create additional solids. A boundary solid grammar uses an initial solid and a set of rules to produce a language of solid models.

Mountain grammar is defined by using a lamina as the initial solid and eight rules to modify the mountain's surface. A rule of the grammar subdivides an existing face and randomly moves the position of the vertices of the face. This produces random variation of the surface of the mountain, while the rules recursively subdivide its faces.

Although this application may initially not sound very useful, it may be quite attractive if the creation of novel material structures and compositions are imagined within structures that are designed to behave in a predefined way. Variations on this technique may be used to produce a wide variety of textures on the surface of the interior of

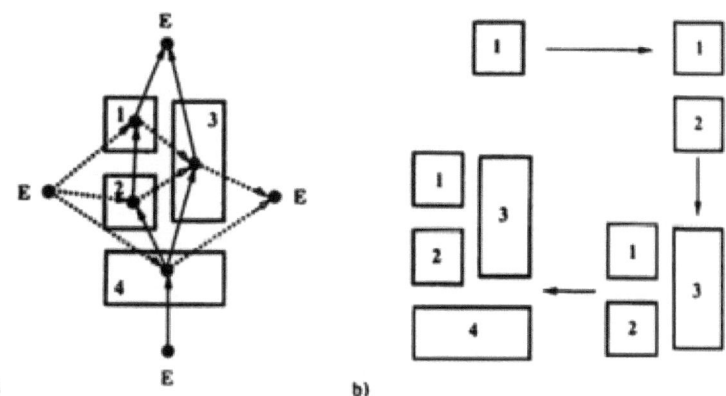

Figure C-2 An example of the LOOS system to define the structure or topology of a layout (Coyne, 1991; reprinted courtesy of Robert F. Coyne).

any solids. Tools such as the ones just described become even more important if you imagine that many of the structures that will be built in the future may be built on a very small scale (e.g., micromachines and nanotechnology).

STRUCTURE SELECTION

Structure selection is a technique for selecting components from a finite list of candidates and ensuring compatibility. Many problems have been formulated with approach in mind and there are a variety of techniques that are used to provide structure to the process and the information needed to perform the task. Many of the early success of expert systems employed structured selection.

TRUTH MAINTENANCE

Truth maintenance is a knowledge-representation technique that records the justification for information so that the fact is removed if the support for a fact is negated or removed. Truth maintenance techniques have been included in several commercial knowledge-engineering tools. The technique is particularly useful in exploring multiple options but has not had the impact that was expected.

Appendix D:

Knowledge-Based Integrated Design System[1]

INTRODUCTION

Engineering design and manufacturing process development are crucial components of the product realization process. They are the means by which new products are conceived, developed, and brought to market. The ability to develop new products of high-quality and low cost that meet customer needs is essential to increasing profitability and national competitiveness. Improving the practice of engineering design and manufacturing is essential to achieving industrial excellence.

Competitiveness demands high-quality products to satisfy customer's performance needs (e.g., ease of final use, expendability, aesthetic appeal, and freedom from defects). Higher quality products require higher quality components and manufacturing processes. Approximately 70 percent or more of the life-cycle costs of a product is determined during design. Fixing defects and errors during design to achieve a quality product is inexpensive. It is much more expensive to fix defects if the customer finds them in the product after delivery.

Today, product realization is a series of sequential activities. During the product design phase, there is a minimum focus on producibility. The ability to manufacture the product is not considered until after the product has been designed. Communication between product design and manufacturing is lacking, and collaboration among subassembly suppliers and part manufacturers (vendors) is rare. As a result, time-to-market is longer than necessary, and final product quality may be poor relative to what it could be.

This section reports a case history that substantiates the use of advanced computer-based technologies and support for workflow process to increase the competitiveness of original equipment manufacturers (OEM) and small-to-medium size supplier enterprises (SME). The results reported are based on an Air Force Manufacturing Science program.

The design system is focused on reducing the design-to-build time for new products. This system would be a knowledge based, integrated design system for helping OEMs and SMEs to build and maintain technological leadership in the world marketplace. The overall mission of such a system is to reduce the time for delivering the first-quality production parts to the marketplace by dramatically reducing the time for product design and manufacturing process development. Such a system would enable *virtual manufacturing organizations* to be easily assembled to provide the support required by the OEM for bringing a successful product to market.

Companies of the future are envisioned to be more like *solar* systems, where OEMs will be surrounded by a plethora of highly efficient SMEs in a flexible network. This flexible network would include banks, community colleges, technology providers, and SME suppliers. This new way of being competitive would still have the advantages of a vertically integrated organization as well as the flexibility and lower overhead of such a network. These flexible networks will be both local and regional and will utilize national server nodes for materials, product design, and process development with access by OEMs and their suppliers provided over the national *electronic superhighway.* .

THE CASE STUDY

The case history is the development of a *Blade Design Assistant* for the Allison Division of General Motors, with the collaboration of IBM and UES, Incorporated, to demonstrate the benefits of an integrated design system that (1) streamlines the workflow; (2) integrates the application tools used in engineering design and manufacturing; and (3) integrates the SME supplier industries with their customers, the OEMs. The Allison Gas Turbine case history was accomplished by applying the methodologies developed under an Air Force manufacturing science program for process design to blade design (i.e., product design at an OEM).

---

[1] Case study supplied by Harold L. Gegel, committee member, as a typical example of the materials selection systems currently available.

APPENDIX D:

The knowledge based integrated design system was designed to have a client-server architecture, where the server was intended to be a massively parallel computer. Using this architecture, a global methodology was developed for designing unit fabrication processes starting at the product specification stage of the product realization process. The design activity was structured as four stages: (1) design clarification, (2) conceptual design, (3) embodiment design, and (4) detailed design.

During the design clarification stage, the functional requirements that will satisfy the customers requirements are established. The functional requirements are then electronically passed to the design team responsible for conceptual design. In this design stage, all of the alternatives for satisfying the are concurrently analyzed to generate a response surface, which then is analyzed to obtain a set of near optimal design parameters. This is then reanalyzed to make certain that this near optimal set of parameters satisfies the functional requirements, which were originally agreed upon by both the OEM and all elements of part manufacturing (including the lower tier tooling vendors).

In the embodiment design stage, the process design may be further optimized by performing parametric studies on the various design parameters. The refined process design is then electronically passed to the detail design stage, where most of the engineering effort is spent.

The design activity that was briefly described above introduces a new design concept called *soft optimization and soft automation*. The conceptual design stage can be automated to a degree by utilizing artificial intelligence techniques such as *neural network analysis* that automatically provide a list of possible process-design alternatives, depending on the functional requirements defined during the design clarification stage.

This approach allows the designer to deal with real-world problems, where the best is only a *theoretical ideal* that is often unattainable or not cost-effective. Through the use of soft optimization techniques a modest goal of being *just good enough* can be achieved even for problems in manufacturing currently considered to be beyond reach *by Calculus-based methods*. This approach to conceptual design reduces the time for arriving at a set of design parameters that will suffice, since design problems are often incomplete (i.e., it is highly unlikely that efficient algorithms for the solution of these problems of arbitrary size will be found).

The design approach used in this research was a component of a total integrated product/process development strategy that requires the simultaneous and integrated development and qualification of all the elements of a total system, as contrasted to a sequential development process. Integrated product/process development requires a two-way flow of information between the customer (the OEM) and the lower tier SME suppliers. This is illustrated in Figure D-1. Integrated product/process development increases the focus on products and processes, improves horizontal communications, establishes clear lines of responsibility, delegates authority, establishes clear interfaces with industry, and changes the acquisition process expectations.

The aim of the knowledge based integrated design system was to develop an advanced process design system. A global design methodology was developed for designing a wide range of unit processes (e.g., casting, forging, extrusion, and sheet metal forming), starting at the product specification stage of the design process (i.e., the stage of the product realization process where the product designers have defined a set of functional requirements for the part).

The program was a team effort. The team consisted of two OEMs, a software developer and system integrator, several vendors, and universities. A structured design process that systematically moves from qualitative to quantitative process definitions was developed. Figure D-2

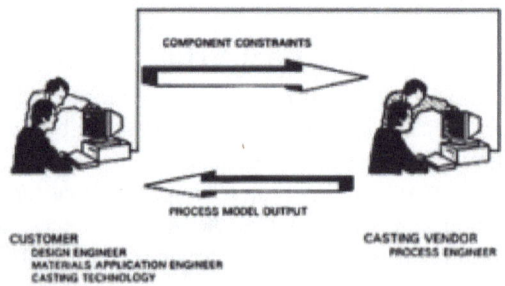

Figure D-1 Information flow in integrated product/process development.

APPENDIX D:

illustrates the *methods developer's* frame of reference for developing and negotiating design criteria. The *product definition*, which passes from the product design activities, communicates across standard interfaces and becomes the initial condition (i.e., the functional requirements) for developing the formal process design definition.

The process definition method consists of procedures and rules (axioms) for each design activity and subactivity to ensure that "what the product designer wants is the same as what the product will possess after processing." The methodology that was implemented was a formal axiomatic design procedure for the creation of synthesized solutions in the form of products, processes, or systems that satisfy perceived needs through the mapping between the functional requirements in the functional domain and the design parameters in the physical domain, through the proper selection of design parameters that satisfy the functional requirements. This mapping process is nonunique, and more than one design may result. Therefore, the concept of soft optimization or being good enough based on heuristic or rule-of-thumb methods of design are emphasized for achieving cost-effective designs.

At the Allison Gas Turbine Division of General Motors, a knowledge based integrated design assistant was created based on flow charts of the workflow process already developed by the customer. In addition, the customer had identified all of the FORTRAN application programs and how they were used in the design process. Before the blade design assistant was developed, the

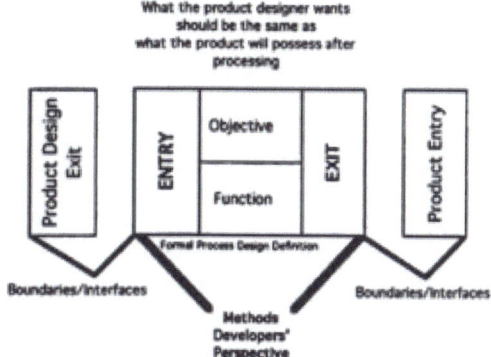

Figure D-2 Methods developer's frame of reference for developing and negotiating criteria.

compressor blades were being designed by a group of engineers using the several workstation and mainframe FORTRAN application programs illustrated in Figure D-3.

The implemented design methodology combines the existing FORTRAN application programs and the turbine blade design process into a blade design assistant. This method assists the engineers in the design of a blade and its associated attachments. The four major engineering roles for blade design are aerodynamics, stress analysis, dynamics, and mechanical design. Individual roles have been created for each of these activities. Within each role, design activities are supported that include design steps, data entry, coordination between other engineers and use of software application programs. None of these FORTRAN programs were altered. The blade design assistant builds the appropriate input parameter file and JCL Stream to invoke each of these applications.

The engineer was not concerned with the format of this file, only the content (i.e., the value of the parameters). Where possible, design activities were performed in parallel; however, dependent activities were prevented from being executed until all requisite information and approvals were available. This procedure was performed to ensure that engineers were not expending effort on inappropriate activities. This procedure enforcement was accomplished using the coordination features to the assistant. The overall goal of this assistant was to shorten the cycle time for blade design and to decrease the effort expended by the design engineers.

Flow charts developed by Allison Gas Turbine were converted to a workflow process model and implemented in the computer via a knowledge-integration shell (i.e., The KI Shell[TM]). The shell development environment was used to create the workflow process model; to implement analytical code to analyze application output and apply design constraints; to prepare input; monitor status; to retrieve output of application on heterogeneous computers; and to suspend or initiate workflow process for different specialists based on the current state of design.

An overview of the method for blade design with the subprocesses for the different roles is given in Figure D-4. The activities associated with each role were grouped or "framed" into *process frames* in different ways. For example, the activities associated with the dynamics engineer role had to be performed in sequence, whereas a choice in the order of activity execution was

APPENDIX D:

allowed for the design review role. Data that was common to all roles was maintained in *information frames* (e.g., material type was data that must be accessed and updated by all roles). Information frames do not have any implied control sequencing.

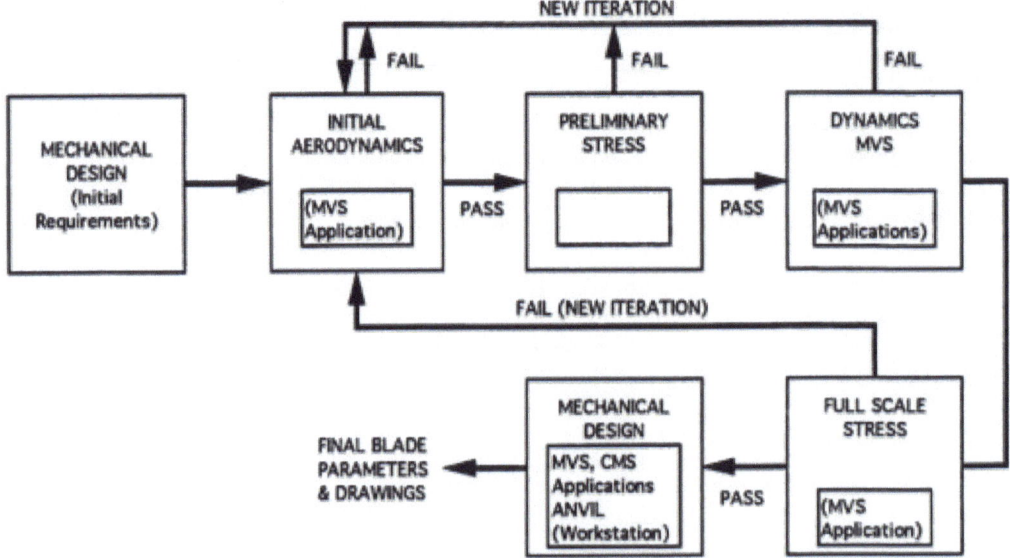

Figure D-3 Control flow between roles in the blade design assistant.

Also associated with each activity are rules that govern the circumstances (i.e., failed design parameter) under which a copy or *instance* of a role was created for another user. There may be many instances of the same role created during one design. In blade design, this was used to try out different design parameters, which were maintained in the database. More generally, as decisions were made during activities, they were maintained in the database.

Finally, the mechanical engineer role in Figure D-4 also illustrates how an activity can in turn require activities in another frame to be executed. The subactivity link was used to link an activity to another frame. Application interfaces were implemented to submit batch jobs via the communications network existing between the workstation and the host.

The KI-Shell features used in the blade design assistant were: (1) multiple roles; (2) roles that create instances or other roles; (3) roles that wait for other roles to finish; (4) ability to execute different roles from multiple workstations; (5) ability to store history of iterations (i.e., process instances and other blade designs); (6) ability to store sets, matrices, etc.; (7) persistent storage of design state and data; and (8) display of 2-dimensional graphs.

The technologies involved in this project include:

- various artificial intelligence methods, genetic algorithms, heuristic and other randomized strategies for soft optimization and automation of the engineering design activity;
- process modeling software that couples heat, fluids, and stress with materials science for predicting microstructure and property evolution during part manufacturing;
- high-performance computing to calculate optimized product design and manufacturing process alternatives;

APPENDIX D:

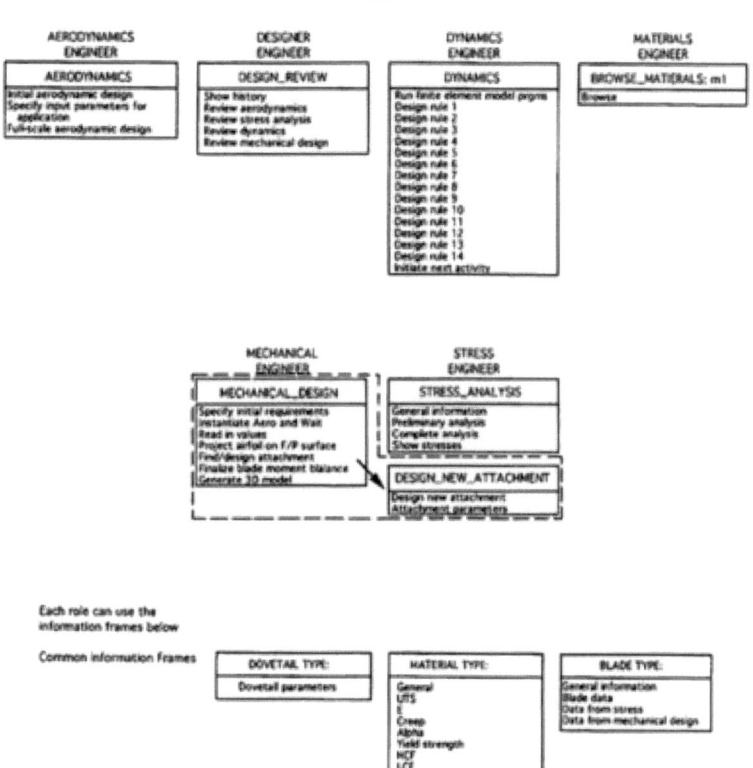

Figure D-4 Blade design assistant.

APPENDIX D:

- high-performance storage systems and communications to move large data files (e.g., modeling results) among storage devices, massively parallel computers, and high-performance workstations;
- hypermedia technology environments that allows a user to discover, retrieve, and display documents and data by clicking on hyperlinks-terms, icons, or images in documents that point to other related documents; and
- material property databases to support process modeling that use the finite element method.

Appendix E:

An Intelligent Knowledge System for Selection of Materials for Critical Aerospace Applications[1]

INTRODUCTION

A key requirement for the successful implementation of the unified life-cycle engineering concept for aerospace structures design is the availability of a well-developed intelligent knowledge system for the selection of materials for specific components (Burte and Harmsworth, 1989). There are two reasons for this: (1) the increasing complexity of the requirements for material performance for any components and (2) the wide range of candidate materials, particularly the newer and more-sophisticated, *advance* materials classes. There are two additional complicating aspects to this part of the problem. First, the newer, high-performance polymers, ceramics, and composites are difficult to identify and compare because of the lack of standard nomenclatures and test procedures. Second, data are becoming available so rapidly on so many materials that the task of keeping a database current is enormous.

A computerized diagnostic program to ensure that all of the important properties and characteristics of all logical candidate materials are considered and that they are analyzed with appropriate priorities would be of great value for reliable material selection. A study demonstrated the technical and economic feasibility of developing a computerized intelligent knowledge system for materials specialists and designers (IKSMAT) in the screening and selection of a wide range of materials for critical aerospace applications (Kaufman, 1988). Further it has been demonstrated that the IKSMAT has the potential to provide great flexibility in query, search, and analysis options, to be very easy for engineers and scientists to use, and to be easily and economically expanded to include many additional applications.

The program described below covered the production of a prototype IKSMAT that provided material-search capabilities for a wide variety of aircraft components.

VISION OF THE SYSTEM

The goal of the program was to develop and build a prototype version of IKSMAT. It was to provide vital guidance in the selection of alloys to meet sophisticated design requirements for spar applications and also form the basis of a system that could be expanded to encompass a broad range of materials (e.g., polymers, ceramics, and composites), components (e.g., engines and empennage), and applications (e.g., helicopters and missiles). The various elements undertaken were

- knowledge base development, data qualification, and interface refinement for the following aerospace components: wingspar, bulkhead, upper wingskin, lower wingskin, fuselage, landing gear, and pivot/swivel fitting;
- programming the system logic, query rules, and response options;
- system design and assembly; and
- establishment of a master database based primarily on MIL-HDBK-5F (DOD, 1986).

TECHNOLOGIES INCLUDED IN THE SYSTEM

Conceptual Model of an Intelligent-Knowledge-System

One generalized model of an IKSMAT applicable to the material selection problem defined above is illustrated in Figure E-1 (Kaufman, 1988). In this model, the knowledge base is the catalog of design and performance criteria for specific structures and the relative importance of the individual criteria in the performance of those structures, which interfaces with the material properties database covering the range of materials and properties of interest. This knowledge base is interfaced with programs permitting users to compile (knowledge acquisition) and utilize (inference engine) knowledge and data to solve problems. The system may be used independently to aid

---

[1] Case study supplied by J. Gilbert Kaufman, committee member, as a typical example of the materials database systems currently available.

APPENDIX E: 64

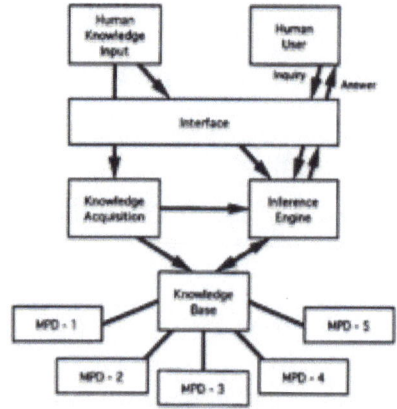

Figure E-1 Model of an IKSMAT applicable to the material selection problem.

in the tracking and selection of materials for specific applications or interfaced directly with the early stages of the design process to illustrate the impact of the utilization of advanced materials on component/vehicle performance.

It is important to note that while the knowledge base itself is the foundation of the system, it is essential to have a well-developed database of reliable, well-documented information on which the inference engine can operate. No matter how sophisticated the logic incorporated into the system, it has little value unless users can have a high-level of confidence in the completeness and quality of the underlying knowledge base (Ambler, 1985; Kaufman, 1986a, Reynard, 1987).

IKSMAT Architecture and Operating Capabilities

The specific system architecture required for the prototype is illustrated in Figure E-2. It is composed of two groups of components, one in the user's facility and the other in a "master database facility." The user interface, controller, and supporting database and knowledge bases are maintained with their own database management system at the user's site. The "master" database facility contains the evaluated data prescribed by materials specialists from the Air Force and aerospace industry through activities such as the MIL-HDBK-5 Coordination Committee and maintained in MIL-HDBK-5 itself (DOD, 1986). It also contains other "external" sources of data, such as those made accessible by the National Materials Property Data Network and the Scientific and Technical Information Network (Kaufman, 1986b).

The user interface handles interactions with the user and provides the display screen control. The controller, the inference engine in the conceptual model discussed above, carries out the expert system functions, most of which will be broadly applicable to other components and other applications. The rules for material selection and design of specific components constitute the knowledge base in this model; extension of the system to handle additional components and applications involves adding to the logic in this knowledge base. Based on information in the knowledge base, the controller passes information to the user interface to determine what actions are required and then interprets the user input to generate queries in the search query generator.

The local database management system deals with the information in "temporary" data files supplied by the user in a manner completely compatible with the permanent databases, and the responses to the user are completely transparent in this respect (i.e., the user will not see the operation as two separate systems interacting), yet the integrity of the permanent databases is maintained. Two user files are kept: the personal database contains user-specific information and while the update database contains data agreed on by the entire user group as "pre-standard."

As noted above, the permanent databases containing evaluated data, such as those based on MIL-HDBK-5, and the principal search software would reside at a remote

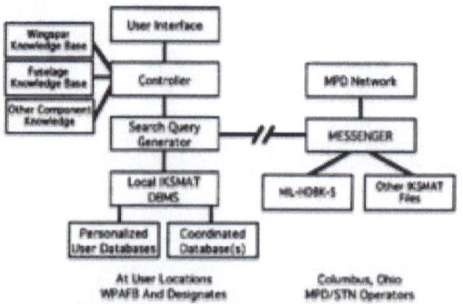

Figure E-2 Specific system architecture for the prototype IKSMAT.

APPENDIX E:

location (the National Materials Property Data Network on Scientific and Technical Network, Columbus, in this case), and all additions and deletions to these databases would be carefully controlled by the appropriate agencies (like the MIL-HDBK-5 Coordination Committee).

The system operates by applying rules based on application or component in the comparison and ranking of individual materials in prescribed sequences, gradually eliminating candidate materials based on their inability to meet stated criteria and the presentation in priority order of surviving candidates. The system must be flexible and dynamic in the sense that new material options may be incorporated at any time and the rules may be altered as necessary to reflect changing vehicular or structural performance requirements.

To be more specific, user approaches to IKSMAT must be of several general types of varying sophistication, including but not necessarily limited to the following:

- user works within existing database and predefined material selection/design criteria and logic to identify optimum candidate materials for specific application;
- user adds new materials to database and then performs analyses based on preset material selection/design criteria and logic to select optimum candidate materials;
- user adds new properties for materials in the knowledge base and then performs analyses based on existing or new criteria involving those new properties to identify candidate materials;
- user redefines priorities associated with existing properties and criteria for material selection and performs new analyses to determine the effect of the changes on the preferred candidate materials;
- user inputs new criteria (specific properties or design-related parameters) and defines their priorities and then carries out analyses to determine candidate materials; or
- user conducts general unstructured search with self-generated queries based on material, property, and parameter criteria.

Functionally, there are several additional features beyond those within the internal IKSMAT logic base and knowledge base that are considered important to a valuable materials information retrieval and analysis system. These are

- ease of understanding for the occasional user who is not an information professional trained in the language and command structure of traditional on-line search systems like Dialogue;
- flexibility with regard to the use of materials nomenclature and property terminology, permitting the user to use any technically correct names or terms (aliases) and still be able to locate the desired information; in addition, the user should have the ability to easily query the system about the meanings of the terms or abbreviations encountered in the process of searching; and
- easy access to many other sources of materials property data, beyond those upon which the programmed materials selection process is based, so that newly generated corroborative or contrasting data may be located, retrieved, and analyzed quickly and efficiently.

The materials to be included in the prototype IKSMAT will include all steels, aluminum alloys, magnesium alloys, and high-temperature alloys included in MIL-HDBK-5. Among the set of candidate criteria for searching for aerospace materials are the following:

- critical crack size index—square of ratio of plane strain fracture toughness to yield strength, an index of the critical crack size at the yield strength;
- stress-corrosion cracking susceptibility—ratio of maximum tensile stress for resistance to tensile yield strength (as an alternative, the ratio of stress intensity threshold for stress-corrosion cracking to plane strain fracture toughness might be used);
- stiffness efficiency—ratio of modulus of elasticity to density;
- tensile or yield strength efficiency—ratio of ultimate tensile or tensile yield strength to density;
- fatigue crack initiation resistance—fatigue strength at one million cycles of life, with stress ratio, R= 0.0, with smooth (K_t =1.0) and notched (K_t =3.0) specimens;

APPENDIX E:

- rate of fatigue crack propagation—fatigue crack growth rate at an applied stress intensity (R= 0.0) of 10 ksi*in**0.5 (preferably based on spectrum loading, but no consistent standard exists); and
- fabricability/cost index including factors such as initial cost per pound; special fabrication requirements like finishing or joining; or multiple sources (materials with high production costs or time lines would have low index numbers).

STATUS OF DEVELOPMENT AND BARRIERS TO IMPLEMENTATION

Knowledge Base/Database Content Development

The content of the knowledge base was established based on the information obtained from synthesis of the guidelines provided by several aerospace designers and finalized in discussions with General Dynamics, Fort Worth. While it was difficult to establish a consensus design approach, once the approach was present it was not difficult to build the associated rules to parallel the analytical process and map the related series of decision criteria.

Compilation of the content data needed for the master database proved to be a much more difficult task than anticipated because of the paucity of reliable, statistically meaningful property data available for any but the simplest of MIL-HDBK-5 design data. For example, even within MIL-HDBK-5, notably in the areas of fatigue, fracture toughness, and stress corrosion (key elements to critical aircraft design), there are very few consistent and statistically based data. This need could also not be satisfied from other sources; most are far less reliable than MIL-HDBK-5 insofar as quality and consistency of data are concerned.

IKSMAT Design

The overall IKSMAT design was satisfactorily completed. The knowledge base and controller designs were practical, and rule implementation was completed, including a strategy for programming ranking logic.

Programming

Programming of the IKSMAT was completed to the point where full-scale interactions with the master IKSMAT database assembled could be tested. Menu interface and presentation formats were also programmed, all to be operational within the National Materials Property Data Network and the Scientific and Technical Information Network, International, MESSENGER mainframe software.

Electronic Data Acquisition and Loading

This, like locating the original data, proved to be one of the most difficult and expensive tasks. Only a partially complete version of MIL-HDBK-5 could be created because of the complexity and variability of data and data formats, even within MIL-HDBK-5. In addition, developing machine-readable updates to MIL-HDBK-5 in a protocol needed to match the master database host in the MESSENGER language on Scientific and Technical Information Network, International, was necessary roughly every six months and proved so expensive as to be prohibitive, because the handbook was at the time produced as a hard-copy product, and the machine-readable updates were generated after the fact. Development of a machine-readable master version of MIL-HDBK-5 will solve this problem.

METHODS TO OVERCOMING BARRIERS

The barriers identified above prevented the production of a wholly satisfactory prototype IKSMAT and precluded any plans to commercialize IKSMAT at that time. The process demonstrated that while it may be possible to conceive, design, and create technically capable and logical artificial intelligence systems for concurrent engineering, the system may be of very limited value because of (1) the limitations of the available data, both in quantity and quality, and (2) the high cost of placing numeric data in useful machine-readable formats for the extensive manipulation needed in such systems.

The actions needed to eliminate these barriers include:

- placement of much more emphasis by government and industry in building reliable, statistically meaningful material property databases that may serve as the foundation of intelligent materials selection and design software;
- maintenance of master versions of materials databases in machine-readable form, readily updatable and readily duplicated and distributed for broad use; and
- utilization of flexible software systems capable of rapid manipulation and varied presentation (e.g., graphical analysis and presentation of complex numeric data complemented with engineering-oriented, intuitive menu-driven interfaces).

APPENDIX E:

Appendix F:
Biographical Sketches of Committee Members

FRANK W. CROSSMAN is director of Material Sciences for the Lockheed Palo Alto Research Laboratory. He received his B.S. from Cornell University and his M.E. and Ph.D. in materials science from Standford University. He is a former member of the National Materials Advisory Board.

JAN D. ACHENBACH is professor and director of the Center for Quality Engineering and Failure Prevention at Northwestern University. He received his Ph.D. from Standford University. He is a member of the National Academy of Sciences and National Academy of Engineering.

HAROLD L. GEGEL is director of the Processing Science Division of Universal Energy Systems in Dayton, Ohio. He received his B.S. from the University of Illinois, Urbana, and his M.S. and Ph.D. in metallurgical engineering from Ohio State University. He was previously a research metallurgist for the Air Force Materials Laboratory.

RICHARD N. HADCOCK is vice president of RNH Associates. He was previously director of advanced development for Grumman Aerospace Corporation. He received his B.S. from the Royal Aeronautical Society in Britain and is a Chartered Engineer.

THOMAS S. KACZMAREK is system architect and program manager, Math Based Process for Dies for the North American Operations Manufacturing Center of General Motors. He received his B.S. from the University of Wisconsin and his M.S. in electrical engineering and Ph.D. in computer and information sciences from the University of Pennsylvania.

J. GILBERT KAUFMAN is vice president of Technology for The Aluminum Association. He was previously President of The National Materials Property Data Network. He received his B.S., C.E., and M.C.E. at Carnegie Mellon University.

MICHAEL ORTIZ is professor of engineering in the engineering department of Brown University. He received his B.S. from the University of Madrid and his M.S. and Ph.D. in civil engineering from the University of California, Berkeley.

FRIEDRICH B. PRINZ is Rodney H. Adams professor of engineering in the departments of mechanical engineering and materials science, Stanford University. He received his Ph.D. in physics from the University of Vienna, Austria.

JAN SCHREURS is a fellow scientist at the Westinghouse Science and Technology Center. He received his Ph.D. in materials science from Northwestern University in Evanston, Illinois.

VOLKER WEISS is professor and chairman of the Department of Mechanical, Aerospace, and Manufacturing Engineering Center at Syracuse University. He is also director of their Engineering Physics Program. He received his B.A. from the Vienna Technical University and M.S. and Ph.D. in solid state sciences and technology from Syracuse University.